MY SURGERY
MY SURGEON

Also by M Clement Hall

Non-Fiction
The Locomotor System—Functional Anatomy
The Locomotor System—Functional Histology
Architecture of Bone; Luschka's Joint
Lessons in Histology
Palestine—The Price of Freedom; Intifada
IME—The Word Book; Independent Medical Examinations
The Fibromyalgia Controversy; Modern Eye Surgery
Aesthetic Surgery; Diabetes Mellitus
Washing Away of Wrongs; A Calendar of Miseries
Murder of Richard Hunne
Pope Innocent III; Saint Benedict of Nursia

Charles River e-books
Arab Spring; History of Afghanistan; History of Syria;
History of Ancient Egypt; History of Modern Egypt; History of Iran;
History of Hamas; History of Hezbollah

History of Charlemagne

Memoir
Viet Nam 1963; Viet Nam 1964-1966; Vale Viet Nam

Fiction
Trauma Surgeon; Spare Parts Box
The King George Inn
Martin's Absolution; Martin in Byzantium
Diamonds in West Africa; Farmer George

MY SURGERY
MY SURGEON

M Clement Hall

MY SURGERY. MY SURGEON

Copyright © 2014 by M. Clement Hall

ISBN: 978-1-304-96764-0

WHY WOULD I NEED TO SEE A SURGEON?

When a patient is told to seek the opinion of a surgeon, have a *consultation* with him, not unnaturally he supposes that an operation is intended or is at least to be considered. I have often talked to patients who told me they had been advised to see a surgeon, a specialist in some particular area of the body, but since they had certainly no intention of having an operation, they decided not to see this surgeon. As it turned out, this decision deprived them of a needed expert opinion on the problem they were suffering.

The fault, if there be one, lies with the medical profession for not explaining itself, and does not lie with the public for failing to understand the doctors' system.

Specialists and specialisation

It is hard to think back to a time when there was no such thing as a *Specialist.* The world has always sought for *the best man in his field* or wanted to know, *who is the top man?* Back in the fifth century BC, the time of Hippocrates, people swarmed to his Greek island of Kos, because they believed he was a better doctor than the man working in their own village, and no doubt he was! Every village has always had a man considered in each and every trade to be better at it than his colleagues, *competitors* is a word we avoid in medical practice! As this man became more busy, because he was so good at what he did, he might take on assistants to do part of the work, and narrow his own activities to the most difficult tasks or the ones that interested him the most.

Diagnosis

The first in the list of reasons for seeing the surgeon is to get his help in finding out what's wrong with you, a process known to doctors as *diagnosis*.

The surgical specialist, otherwise known as the surgical consultant, is in many fields of medical practice the person best qualified to make a diagnosis, in other words to tell you what's wrong with you! Although it doesn't always work out this way, the science behind medical treatment requires that the doctor should have a pretty clear idea what's wrong with you before he embarks on treating you. Anyone with a car will understand that buying a new battery when the motor won't start, isn't going to change the problem if it's really the ignition wires that need to be replaced, and although the battery you've got may be old it is still doing a perfectly good job.

Treatment

Once what's wrong with you has been determined, that is, your problem has been *diagnosed*, the surgeon will then discuss with you his views on the best way of helping to cure, or at least alleviate, your problem, a process known as *treatment*. We emphasise always that **diagnosis must come before treatment**.

HOW DO I FIND A SURGEON?

Family physician referral

In most instances, your family physician will give you the name of a surgeon. Depending on that doctor's own way of conducting his practice, his secretary may make the appointment for you, or you may be given the surgeon's telephone number and advised to make the appointment yourself. Possibly you will not know that a letter will go from your own doctor to this specialist, outlining the details of your problem, as your family physician understands them, the results of the tests that have been made, and the specific reason you are being sent to this surgeon. The letter might be as short as "Lump in left groin, ?hernia ?operation?" or it might be several pages long, and accompanied by multiple copies of X-ray, laboratory and other reports.

The surgeon certainly welcomes all the information he can get about you. It will help him to make the right decision in regard to how he should treat you. It may save him from having to ask you to make another visit while he tries to obtain the information he needs.

Some family physicians will ask you if you have ever been to a surgeon before. If your doctor is as aware of your past medical history as ideally the profession would expect him to be, his documents will show him that. He may ask if you would wish to see the same surgeon again, and make your appointment accordingly. If you have never been to a surgeon before, or if you wish, for one of many possible reasons to see a different surgeon, he may ask if you have some particular surgeon in your mind whose opinion you would like to have.

It is wise to ask your doctor's advice in regard to whether the surgeon who operated on you previously, or the one whose name you got from a friend, is really the surgeon your doctor would choose for you. Surgery has become so specialised that although the surgeon you know might be able to do your operation adequately, your own family doctor might be aware of another surgeon who is more experienced in this type of work, is using a different type of equipment that is for some reason more satisfactory, or who for various personal reasons might make a better "match" with you personally.

When your doctor tells you that he wishes you to see a particular surgeon, it is quite appropriate for you to ask him to explain his choice. You will of course do that in a non-hostile, non-doubting manner, leaving him fully confident that you trust his judgement, but would merely like to be informed about the reason for choosing this particular surgeon.

Happy customers

The decision to have an operation is one that is often reached quite slowly. The majority of operations are "elective," that is, they are performed at the patient's time of reasonable choosing, and when he has reached a point that he has decided his apprehension and dislike of the discomfort and inconvenience of surgery is now less than the discomfort and inconvenience caused by the condition from which he suffers.

He may have been told by his family doctor, "You really ought to get that hernia taken care of," or he may have been told, "The only way you're going to get rid of that pain is to have your gall bladder out," but he hasn't quite made up his mind to do anything about it.

Then one day he meets a friend or hears someone at his workplace who starts talking about having had the same problem. This person talks about putting up with the problem for months or years. He talks about burying his head in the sand and pretending he wasn't really an ostrich. He admits that finally it was his wife who told him she was sick of listening to his complaints and she'd insisted he get an appointment with a surgeon. To cut short this often very long tale, the surgery wasn't nearly as bad as he'd expected, the surgeon seemed like quite a nice person really, not at all what you might expect, and now his problem's gone away and he's had to find something else to complain to his wife about!

You may ask the friend one or two questions indirectly. You might even go so far as to admit you have the same problem yourself. You probably won't admit to having had the same fears. You do decide that, "Here's a happy customer" and it sounds like you ought to go and see the same surgeon. Perhaps, after all, the treatment will be less bothersome than the disease.

So, when you go back to see your family doctor, you tell him you've decided that you really would like an operation, and this is the man you'd like to have do it.

Suggested by golf club members

In choosing your surgeon I distinguish between getting a name from a "happy customer" and hearing a name from an acquaintance, for instance when playing golf, or more typically in the bar after you've done all the holes. This man may of course be a "happy customer." Or he might be the average club blow-hard who knows everything about everything. He can tell you why your car is the very one you shouldn't have bought. He can tell you where you should have put your money in the stock market. He can give you a list of all the surgeons in town, "I wouldn't send my dog to."

His advice on surgeons is as well informed as his advice on the stock market. If you have surgery in mind, avoid this fellow like the plague!

WHAT SHOULD I TELL THE SURGEON?

History of current problem

Every doctor you will meet has his own personal way of dealing with his patients. Some will handle the whole encounter on a stylised, but organised question and answer basis. Some will leave you, the patient, to do all the talking, and you may be left wondering how much they've heard of what you've told them. Some will breeze in, tell you they've, "Got all the information" they need from your family doctor, and as he leaves the consulting room advises you over his shoulder that his secretary, "Will be in touch regarding the date for your operation."

There are arguments for and against any technique of dealing with patients and their problems, but ideally you will have some kind of conversation with your surgeon over the matter that troubles you, and how he would propose dealing with it, and you will be able to ask him any appropriate questions.

Don't suppose your family doctor has told him everything about you

In bygone days no specialist could or would accept a patient "off the street." He was permitted by the code of his profession to see only those patients referred to him, that is directed, to him by their family physician. In his turn the *referring physician*, for which read family doctor, or general practitioner as he was known in those days, was expected to send a letter explaining why he was requesting the

12

specialist's advice.

There were in this method as in all others, some good and some bad points. However the system is no longer followed rigidly. I recall when working in England seeing, literally, a corner torn from a newspaper with "please see" written on it. That was the "referral letter!"

Speaking as a specialist surgeon, I greatly value the assistance of the referring family physician in telling me his reason for sending the patient to me. A note a few lines long is usually adequate, with copies of relevant documents on tests performed, previous medications and so on. One of these days each patient will have a "medical passport" with all of his details on it. Progress in computerisation of data is bringing us near to that possibility. But at the moment information on a patient's medical problems can come only from one member of the profession to another. We have no way of gaining electronic access to information.

But you should not take for granted that the surgeon you are now seeing has received full information from your family physician. He may have received no information at all, because none was sent. Or because, although it was sent, it hasn't yet reached the surgeon, it might be sitting somewhere in the bowels of the post office sorting system; it might still be on the desk of the family physician's overworked secretary. There are any number of "mights." The other possibility, potentially a bigger problem, is that the surgeon has received some of the desired information, not all of it, but he doesn't know which parts are missing. When I have no information at all, I can start from scratch. When I have a lot of information, I don't know about what is still missing. Or put another way, "you don't know what you don't know."

It is wise for both you as the patient, and the surgeon you are discussing your problem with, to start from the beginning, to use any information he has been provided with, as a safeguard and assistance, but to presume that everything needs to be touched on, even if not recounted in full detail.

Your surgeon will want to know in the simplest terms, "Why have you come to see me?" Medical students and residents in training used the symbol *c/o* indicating "complaining of." It is more fashionable now to use the "S" of an SOAP system, the *S* standing for *symptoms*. That's all very complicated, but you may see the doctor writing "S" and now you know why.

Most patients don't like to be asked if they're "complaining." They don't think of themselves as complaining, they'd just like some relief from their problem. Since doctors are still arguing among themselves about what exactly constitutes a "symptom" that doesn't help very much either.

It's just a great deal easier to answer the surgeon's spoken (or unspoken) question, "Why have you come to see me?" And that you can readily do. "My stomach hurts after I eat, and I'd like to be free of the pain," or "my hip hurts when I walk, and I'd like something done about it."

There's no substitute for plain simple language to express a simple problem and a plain wish to have it relieved.

What investigations have been made?

It is common for a patient to have only "treatment" in mind. He knows he has a problem. He wants something done about it. He doesn't understand that although it is all too easy to go headlong into treatment, the result of this treatment is more likely to be satisfactory if a correct diagnosis came first. The surgeon is always under pressure to "do something," and to "do it now." But the surgeon likes this "something" to be the "right thing." And the "right thing" in treatment requires a correct diagnosis.

To make a very simple comparison, if you have an electric short-circuit somewhere in your house, the electrician could keep changing parts of the system hoping blind luck will eventually lead him to change the defective part; alternatively he could examine carefully and systematically the components of your house electrical system to determine where the defect lies, and then repair or replace only the defective portion.

I have the greatest respect for electricians who are able to do this. I have also the greatest respect for surgeons who decline to be pressured into performing the "obvious" operation, and prefer instead to set about the necessary tests to be sure they have the correct diagnosis.

Only the replacement of the defective electrical part will result in the correction of the electrical fault, and only the correction of the defect in the human problem will relieve the patient's situation. All the electrical parts may be well used and ageing. All the body parts may be showing the effects of time, but if the electrician or the surgeon does not focus on the part that matters, then the problem remains uncorrected. A correct diagnosis is essential to appropriate and successful treatment.

What treatment has already been given?
Your family doctor may in his letter of referral have mentioned the medication that had been used, and he may have mentioned the operative procedures already tried. But if you want to help your surgeon to understand your problem fully, if you want in fact to help yourself, you will write down ahead of time everything that has already been done in an attempt at relieving this problem.

If your problem is of long standing, or if you have for one reason or another changed your family doctor, it is possible that your family doctor does not know all of the previous treatment and investigations that have already been made. Today's mobile urban society, and the free access to drop-in clinics, hospital emergency departments effectively operated as non-emergency drop-in clinics, and house call services, results often in the family physician, if there even is such a person, having at best an incomplete and often inadequate record of his patient's previous health care. In such circumstances, your surgeon can only obtain this information from you. A little thought on your part, anticipating questions like, "which doctor did you see?" or "where and when was that operation performed?" will help your consultant surgeon to get a more complete picture of what has gone on in your medical past. You shouldn't present him with, effectively, an "if it matters to you, find out for yourself" attitude. He's trying to help you. It's you it matters to, not him!

What was the effect of this treatment?
If you have had medication prescribed, did it relieve the problem?

- Was the relief long-lasting or only briefly helpful?
- Even though it was helpful, was the quantity prescribed used up, and you didn't get a refill on the prescription?
- If it was of no benefit, did you tell the doctor this, and was any other medication prescribed?
- Or, if it was not helpful, did you just say nothing about it?
- Did the medication cause unacceptable side-effects? Was that the reason it was discontinued?
- If you have had an operation, did it relieve your problem? If so, for how long?
- Did other new problems result from this operation? What was done about them?

To the extent you can, bring all documents with you that will fill out the details of any investigations, operations and other treatment. If for instance, the operation was on your stomach, the bare statement, "a stomach operation," tells your surgeon very little of what he needs to know; he will have to have much more detailed information than that before he can make any serious decision regarding further treatment.

"Ownership" of records

The law on ownership of medical records is quite definite, but surprises the patient when he learns it. Doctors, hospitals, and other institutions where medical care is rendered have a legal obligation to maintain records of that treatment. For this reason the actual physical document, the pieces of paper on which the treatment is recorded, is considered the property of the institution where the treatment was carried out. They cannot with any safety to themselves give away their documents. A family doctor will sometimes give a patient his records to take to a specialist when the patient goes for a consultation, but that family doctor is putting himself at some risk since he has an obligation to maintain that record for a specified number of years, he does not transfer that obligation when he gives his records to someone else.

The right that you do have as a patient is to the *information* in the record about yourself, not to the physical record itself, but to the information in the record. Effectively that means you have the right to view the actual record, or to have a copy of the record made for you.

Most doctors' offices, and all hospital medical record libraries (the part of the hospital where your record is stored), have copying machines. Doctors occasionally experience a troubled or troublesome patient who walks in at the most busy time, and demands to be allowed to inspect his records. The patient has the right to see his record, but he must do this at reasonable convenience to everyone else. The usual practice is to make a copy of the record and to give the copy to the patient. For this service a "copying charge" may be made. This covers the time spent in searching for the record, the doctor's time in looking through it to make sure it is appropriate, and secretarial time in doing the actual copying, not to mention use of equipment and materials. The doctor has to pay for this out of his pocket, the government gives him neither the secretary nor the equipment!

Whether the family doctor has the right to provide copies of other doctors' reports is a point argued in law. When it is a simple matter of seeing that the patient gets appropriate treatment, there usually isn't much concern. When it becomes a matter of copying a patient's records for court or other legal purposes, then that raises an issue that it is best to leave to the lawyers to determine.

If you go to the hospital to get copies of your records, you may find that they charge you a great deal more than your doctor cares to do. This is particularly likely to happen if the records are requested for legal purposes, and extensive copying is required. The charge levied, however, has no bearing on the expenditure of the medical record clerk's time or the hospital's materials. It is related to the hospital's need to improve its finances, so that a $5 charge from your doctor's office might well be a $50 charge from the hospital. Caution is advised in requesting copies of records from a hospital, find out ahead of time what they are going to charge you!

X-rays

"Reading" an X-ray is often a matter of judgement. The radiologist whose business it is to make reports on every X-ray, sees only the X-ray picture, and generally has no personal contact with the patient. Although he's supposed to have been given a full explanation for the reason the X-ray was ordered, he often doesn't have this, or it isn't to hand at the time he makes his report. Your surgeon, who will see both you and the X-ray, is likely going to want you to bring the actual X-rays with you, to his office, when you have your consultation.

This requires that you go to the hospital or clinic where your X-rays were made, and ask them in the X-ray department (now usually called the "imaging" department) to let you have your X-rays so you can take them with you, for your consultation with the surgeon.

Once again you are likely to run into problems of "ownership." Legally, the X-ray picture is the property of the hospital or clinic where it was made; the record to which you have legal access is not the picture, but is the radiologist's report. The radiologist, like your family doctor and surgeon has the same obligation to store the paper records of the reports, but is not under an obligation to store the picture, usually called a "film" or in doctor-talk, "the X-ray." I have been given two explanations for this. Firstly the films can be extremely bulky, and storage space is at a premium. Secondly, the film has silver emulsion on it, which has value, so the hospital or clinic likes to retrieve this value, rather than leave it buried in a storage cellar.

The end-result of this is that if your examination was made recently in a small clinic or small hospital, they can probably find your X-rays and will let you take them away, requiring you to promise on your mother's grave that you will bring them back immediately they are no longer needed. It helps them, and hence it helps you, if you phone a day or so ahead of time to say that you want to pick up your X-rays. They may well have been put in the basement storage room, and there may be no one available in the department to go and fetch them if you don't give them notice that you would like to have them to take to your surgeon; they will usually ask you to whom you are taking them.

A problem generally comes when you are dealing with a large hospital. There seems to be a rule in hospitals that the largest and most famous hospitals are the ones least likely to be able to find anything. They often are also the most financially grasping. They may refuse to let you have the originals of your X-rays and instead demand a sum of money, upwards of $100 to provide copies of the X-rays; these copies are then your property to keep. Check them out before you get caught up in this!

As an orthopaedic surgeon, it is probable that nearly every patient I see will need to have an X-ray; most have already had their X-rays taken before their family doctor sends them to me. I'm never sure whether the patient believes this himself, but a common answer to my question, "Did you bring your X-rays?" is to be told, "My doctor says he was sending them." Family doctors don't have the X-rays in their offices, the X-rays remain in the clinic or hospital department where they were made. To me it seems not unreasonable that the patient should participate in his own consultation to the extent of going and getting them himself, or having a relation fetch them on his behalf.

It's a simpler matter in many of the countries of continental Europe and of South America. There they regard the taking of an X-ray as a service. You pay to have the X-ray made, they take the picture and give it to you, just like a photographer. The advantage of this system is that you can always keep your own picture. The disadvantage is that most people lose their pictures and then there is no record from the past, that would otherwise have been available for the guidance of the doctor in the future care of his patient.

Times are changing in "imaging." Much is now recorded "digitally" (ask your child to explain that term!). Film for X-rays is going out of fashion, the image is now made on a computer disc, and if your specialist is going to view this he will have to have a computer in his office. The younger ones will, the older ones might not, although being "tech'd" is by no means tied directly to age.

History of previous problems

You should come prepared with dates and details of all past medical events. Remember always that your surgeon is asking you these questions for your own benefit, he's trying to help you. In return you can best help yourself by a little preparatory work. There are some persons, usually the young, who have never in their lives had any significant medical events. But most of us have had some operation or injury of consequence, or continuing medical problem. Work out in your own mind when these happened, within a few years if that's the best you can do. Then write them out in order of their occurrence so you can give your surgeon a simple, correct and concise answer to his questions.

The surgeon in taking a "past medical history" is not prying into your personal affairs, he wants to know of any event that might affect your care or be relevant to your diagnosis. Although having had your gall-bladder out may seem to have nothing to do with your forthcoming hip operation, having had your cataracts operated on might be relevant to your ability to get around after the surgery. The patient is not in a good position to judge what is relevant to diagnosis and treatment. It is therefore usual practice for doctors to record all major events, then they are there in the record at any future time it needs to be consulted.

Current chronic medical problems

Your surgeon will need to know about any medical problems that you currently or chronically suffer. "Chronic" is an ill-defined period of time, but generally implies something that has been going on for months or years, and doesn't look like it intends to go away.

Most chronic medical conditions have very little effect on the actual surgery, the "cutting" part of the whole procedure. They may, however, have a very significant effect on your ability to tolerate an anaesthetic. And, remember always, it's not just a cutting operation that might be under consideration, it's also likely to be an anaesthetic.

As I repeatedly emphasise, there is no such thing as "risk-free" surgery or risk-free anaesthesia. But prudent care minimises the risks to the greatest extent possible. Remember always it is **you who is taking the risk**, not the surgeon and the anaesthetist. You, the patient, must assist them by full and free disclosure of any medical problems that have in any way jeopardised your health in the past (even if you think you have fully recovered), or any problems that continue to show symptoms to the present time.

Surgeons, and more particularly anaesthetists, are always amazed at the patient who thinks he's "putting one over on the doctor." Typically, he's well educated, possibly a university professor. He denies having any heart problems. He denies ever having had any heart problems in the past. Only when a routine cardiogram is performed and it is found that there's evidence of major heart disease, does this patient reveal that six months ago he spent two weeks in intensive care for a coronary artery obstruction. When you ask this educated man why he lied, he'll tell you he was worried you'd cancel his operation if he told the truth! If his anaesthetic had not been cancelled he might well be dead, almost a form of suicide.

So don't try and "put one over" on your surgeon. Tell him the truth, the full truth. His enquiries will seem very general as he runs through what doctors call the body's "systems" and will follow a pattern such as "Have you ever had any problems with your heart, blood pressure, lungs, breathing, stomach, waterworks?" and so on. Listen to him carefully. Be prepared ahead of time with your answers. Be prepared with details, dates, and information regarding who treated you, where, how, and with what success.

Insurance and legal consequences

Because orthopaedic surgeons treat persons who were injured in accidents, they are among the most likely of the surgeons to become involved in legal issues, but all surgeons are likely from time to time to find that the patient presents forms to them, requesting support for a claim of short or long term disability. If it is your plan to involve your surgeon in your legal or insurance problems, you should tell him this in a straightforward way at the outset, or better still, tell his receptionist at the time the first appointment is arranged.

If the surgeon finds at a later date that you have not been up-front with him, he may feel he has reason to doubt everything else you have told him. Instead of assisting you to obtain whatever you are rightfully entitled to receive, you may find he avoids involvement in these issues.

By being less than fully open with your surgeon, or any other doctor at any time, you are causing difficulties for yourself. Be open and be honest. You'll find you get more support that way from all of your doctors.

Current medications

Although some persons rarely if ever take any medication, a substantial number of the persons a surgeon will see in his office are taking multiple different drugs. They frequently are unable to list these drugs, and frequently have only a vague idea why they do take them. When the actual medications are examined it is not uncommon to find that the patient goes to different doctors at different times, and that some of his medications are duplicated, having been ordered for the same purpose but by different doctors.

Unfortunately our society has yet to evolve an organised way of monitoring the medications that are prescribed for patients. Certainly this has to do with liberty, democracy, freedom and all the rest of it, but the lack of such a system does not meet the needs of the elderly patient who doesn't speak English, and just accepts whatever medication is prescribed for him. Let us hope that you are not in that group.

Patients on multiple medications are most strongly advised to write down the names from the bottles of all pills that they are taking, and keep this list with them whenever they are away from home. Like the advice your mother gave you, "always wear clean underclothes, you never know when you'll end up in hospital," the same applies to your list of medications. There are some of the larger drug store chains that will perform this as a service, keeping a list of the medications any individual takes. This is very useful, but clearly also has its commercial benefits to the drug store. Other pharmacies will give you a printed label, the duplicate of the label put on the outside of the container; these can be kept to show to a nurse or doctor in the future.

It will help your surgeon if you bring all of the medications you are presently taking, not just a list of them. It has been my experience that lists are often illegible, they may have been written by another person, and they rarely include the instructions, such as "one a day" or "take as necessary."

If you are taking a lot of different medications, it will also help your surgeon if you separate the bottles into those that are:

- regularly taken,
- occasionally taken,
- pills you did take, but no longer need to take.

It is particularly important for your surgeon to know if you were ever on any type of medication that contained cortisone (aka prednisone), and you will be asked this same question by several persons as you proceed through the "chain" prior to any operation, or other procedure performed in the hospital.

Blood thinners and aspirin

It is also particularly important for the surgeon to know whether you have been on any kind of "blood thinner." You should not forget to tell him about regular scheduled use of aspirin, known also as ASA (acetyl salicylic acid), and commonly ordered for persons who have arthritis, or for persons at risk of heart artery (coronary) blockage. Aspirin has a blood "thinning" effect as well as the pain relieving effect for which it is most commonly used. Non-steroidal anti inflammatories (NSAIDs) fall into the same grouping of "blood thinners."

Birth control pills

Most people don't think of birth control pills as medication, but they do come under the same government control agencies as all other medications that require a doctor's prescription. The issue of consequence in regard to your surgery is that persons who use birth control pills are at a higher risk of getting a blood clot in the veins, that then passes to their lungs (known as a pulmonary embolus) and this may cause their death. The likelihood of this occurring is believed to be increased in those who smoke and in less-young women who are still taking "the pill." Your surgeon should be told if you are taking birth control pills. Mechanical birth control devices are of consequence only if you are expecting to have abdominal surgery, unless they have been associated with some infection, in which case your surgeon should be informed.

Allergies

At every point in the "chain" leading up to any surgery, you will be asked if you suffer from any allergies.

Distinction is made between:

a) allergies to medications,

b) allergies to medical equipment and supplies, e.g. bandaids, latex rubber, surgical soaps, iodine and

c) allergies to food and other substances.

a) *Allergies to medications*: Problems are caused in a patient's treatment when they declare themselves to be allergic to just about every medication known to mankind, in particular every antibiotic that's been invented. These lists of allergies seem sometimes to be presented as a means of showing, "I am not as other men." Sometimes they are presented almost as a challenge, "You can't use on me any of the drugs you would give ordinary people, so what are you going to do about it?" Many doctors will politely decline the challenge, tell the patient that their problems are more than they as a doctor feel able to cope with, and suggest they go elsewhere for their treatment.

There are of course persons who suffer from very real allergic problems, and every doctor is aware that an allergy to a medication or other substance precludes him from using that medication or substance. The loss is not the doctor's, it's the patient's. There is a misconception that any one medication could just as well be replaced by another. This is simply not true, there is not always an equally effective substitute. More than that, medications often come in groups or families, so that if you tell the doctor that you are allergic to antibiotic "A" and so cannot take it, this may also prevent him, from using antibiotics "B" and "C."

The term "allergic" has a very definite medical meaning, but regrettably "allergy" is frequently misused, so the patient goes through life labelled as having an allergy which may in fact not be true. Most experienced doctors at some time in their life have encountered a patient who suddenly can't breathe after they've been given an injection of, for instance, penicillin. This is a life-threatening situation. No one would ever suggest otherwise. But every doctor has also met a patient who believes they are allergic to penicillin, and on being asked to explain what happened, remembers that they once took a pill, they're not sure what it was but they had a cold at the time, and three days later they had a headache. This patient, of course, might be allergic to penicillin, but the evidence that they are, is simply not there. However, by their simple statement that they have such an allergy, the doctor is prevented from giving them penicillin and all the other antibiotics associated with the penicillin group.

If you have had some adverse reaction after taking medication, get skilled assistance from a doctor who is prepared to discuss with you in some depth whether the reaction was a real allergy, or whether it was a known side-effect of the drug that did not constitute an allergy, or whether the experience you had, such as a headache three days after taking the medication, had absolutely nothing to do with the medication, and was neither allergy nor side-effect. Then be prepared to tell this to your surgeon, or other person who enquires into "any history of allergies?" exactly what it was that happened to you, who caused a diagnosis of allergy to be made, and why .

It is difficult for both patient and doctor to understand this, but previous use of medication without associated problems does not guarantee that there will be not be any in the future. A genuine allergy to a medication may be found even when previous use did not provoke one.

b) *allergies to medical equipment and supplies*: There are a number of substances and materials that are in common use in hospitals, but to which some people have genuine allergies or abnormal sensitivities.

Iodine is a very effective antiseptic and was routinely used to cleanse the skin before surgery. So many persons are allergic to it that it is rarely used in its pure form now, and even its modified "non-allergic" form is less commonly employed than a very few years ago.

Band-aids will in some unfortunate persons cause blistering of the skin. So will rubber elasticised bandages. There are many other substances that will do the same; unfortunately the patient doesn't know about them until they are used, and the first time they are used on them may be at the time of their operation.

c) *allergies to food and other substances*: There are unfortunate persons who react to foods such as shellfish or peanuts just as dramatically as has been described for penicillin, experiencing sometimes a life-threatening situation. This has no bearing on your surgical treatment, other than creating a concern for an "allergy-prone" state if such a thing exists. It does affect what they will give you to eat in hospital, so if you are staying there long enough to get fed, they will ask you about this.

Other current treatments

You will be asked whether you are currently receiving any form of non-medication treatment. Your surgeon will wish to know if you have been, or are currently, receiving physiotherapy, chiropractic or similar treatment.

YOUR CONSULTATION WITH THE SURGEON

PRELIMINARIES

Appointment time You will be given a time to come to the surgeon's office. This is the time when he would hope to see you. It does not allow for delays in traffic, nor for time taken to register on arrival. It probably makes no difference to your schedule if you deliberately come ten minutes earlier than the time you are given, but it makes quite a difference to the surgeon's schedule if everyone comes to his office ten minutes late.You may notice that I said "the time when he would hope to see you." Surgeons more than most other members of the medical profession are unable to control their day, or night! If he has not previously met the patient who came before you, he can't be blamed for not knowing that her problem was far more extensive than he had been led to believe; or that she was so poor in remembering essential details of her history that it took twice as long as expected to get it from her; or that her English was extremely hesitant, her husband's was no better, and the difficulty in communication had greatly lengthened the interview. Your surgeon may have been called away from the office because of an emergency in the operating room, or the emergency department, his scheduled operations may have started late because of an emergency procedure to which he or someone else had to give priority in the operating room, or they may have taken longer than anyone could have foreseen. The one possible reason for keeping you waiting that is *not* going to be true, is that he is doing it deliberately.

So the "unfairness" of the situation is that you will be expected to come in good time for your appointment, but your surgeon might be prevented, through no choice of his own, from doing the same.

Language

Are you fluent in the language your surgeon will use? If not, you should arrange for someone to accompany you who can act capably as an interpreter. You should recognise that this interpreter may be needed to help during the physical examination, so it should be a person in whose presence you will not be embarrassed during a physical examination.

The surgeon's name may suggest that he speaks the same language that you do. Don't take that for granted; if you wish to act on that possibility phone his office ahead of time and find if he is in fact *fluent* in your language, or the dialect variation of that language which you normally use, and in which you feel most comfortable expressing yourself.

The receptionist

On arriving in the surgeon's office you will let the receptionist know who you are, and you will be asked all the usual personal identifying details of your name, address, telephone number, date of birth, family physician, or the referring doctor if not your family physician, place and type of work. The receptionist, although an essential and important member of the surgeon's team, will not be in a position to answer any medical questions. It is possible that you will later find that the receptionist is in fact also the surgeon's nurse and she may play a significant role in advising you if you have surgery, not the actual operation details perhaps, but all of the other aspects that surround it.

Insurance card

Bring your insurance cards with you; some persons claim they can remember multiple digit identifying codes, some of them do, many of them can't. You may not realise there are other essential details on the card such as "version codes" and because of difficulties with the government health agencies, the receptionist may tell you that they cannot proceed further with your appointment unless you present the card, with the alternative of paying out of your own pocket for the visit, which almost certainly you don't wish to do.

If you have other relevant insurance numbers, such as workers' compensation, extended health care coverage and so on, you should bring those details as well.

Let the receptionist know about special equipment needs, such as crutches, wheelchairs, apparatus for sleep apnoea, devices for insulin injection and so on.

THE INTERVIEW

If you have read the preceding, then very little of what the surgeon may ask you will come as a surprise. He may have a structured and formal way of conducting his interview, or it may seem less formal, but will probably be directed at obtaining the same information.

The structured interview is directed at the following points:

- Nature of your presenting problem.
- History of your presenting problem.
- History of other relevant conditions.
- History of all other previous medical conditions, including illnesses, medications, operations, allergies and so on, which may seem to you to be totally without relevance.
- A general check list going through the different body systems such as heart, lungs, abdomen, spine and joints and so on, to make sure no significant or relevant portion of the medical history has been ignored, known to doctors as a systems enquiry.
- Family history, to determine whether there is anything in the medical background of your parents, grandparents and so on, that might be relevant to your own problem, its diagnosis and treatment.
- Social history, often this term is used only to determine how much you smoke and how much you use alcohol.

Occasionally, when relevant, it might be an enquiry into the abuse of "recreational" drugs. The surgeon may also feel the need to know about your personal living arrangements, what will you do for money when unable to work, who will care for your children and so on.

- There may also be features in the presentation of your physical problems that suggest to him a contributing psychological feature associated with social difficulties, and you may find him enquiring into this. Help him! Don't obstruct him! He's not asking these questions for his amusement, **he's trying to help you!**

- Employment, although this is a routine enquiry, one probably made by the receptionist, the surgeon needs sometimes to have a clear idea of the type of work you do. The job title is of less importance than the actual physical requirements of the job, the loads you lift, the frequency of movements, the hours put in, the possibility of job variation and so on. This will be particularly relevant if you are making a medical claim that relates to your work.

THE PHYSICAL EXAMINATION

Do I have to be examined?

This is a question sometimes put to the surgeon. Some patients believe that the laboratory results, the X-rays, the CAT scan, the reports from other doctors, should all be enough for the surgeon to decide what operation he's going to do, and further physical examination is simply not necessary.

I have full sympathy for this point of view! I personally detest being examined by any doctor, and I believe that most of my colleagues share this same feeling. However, your surgeon may well be under a legal obligation to perform an examination himself, one he cannot put aside without jeopardising his own career.

You are at full liberty to refuse a physical examination, like they say, "it's a free country." But your surgeon will probably tell you that he cannot undertake your treatment unless you permit him to examine you. This is a situation that should never take place and the so-called "Mexican stand-off" should have been prevented by the referring physician advising you what to expect when you see the surgeon. That doesn't always happen, the reason of course why this book has been written.

Chaperones

You probably haven't seen this surgeon before. You may feel uncomfortable being in the presence of a stranger with many of your clothes removed, even if you are wearing that skimpy examination gown. Some surgeons, especially gynaecologists, have for their own protection a female member of their staff present during the physical examination.

No surgeon is likely to object if you say you would prefer to have a relation or friend present during the interview and physical examination. It should be understood that this person is only there to give you emotional support and is not there to participate in any other way. Do *not* bring someone with you who is going to try to tell the surgeon your history, or someone who thinks it is her job to tell the surgeon what type of operation he should perform. Even if your choice of chaperone is present, the surgeon may still choose to have a member of his staff in the examining room, you must understand that there have been too many false accusations made against surgeons for him not to think of the need to protect himself.

Family members

If the family member is present to act as a chaperone, the surgeon is not likely to object. If the family member is present as an interpreter, the surgeon is almost certain to welcome their assistance. But how does he feel about the wife who insists on being present for what was intended to be an interview with her husband, but she answers all the questions? Or the husband who treats his wife like a farm animal and won't let her make any decisions?

These are all situations that each surgeon must decide for himself, occasion by occasion. From the point of view of my own practice habits, it is my belief that each patient is an individual, not a part of a duo. I prefer, other things being equal, to have a "one-on-one" conversation with my patient. I believe in that way they are free to tell me what they think, rather than what they believe their spouse finds acceptable. I can ask questions that they might be reluctant to answer in their spouse's presence.

After the interview and examination, after the patient has begun to think for themselves what treatment they would like to have, I may then invite the spouse in, describe my findings and conclusions and invite any input they may wish to make.

Juveniles

In most jurisdictions it is required that a responsible adult be present if any intimate examination or any treatment of a juvenile is undertaken. It is certainly required that the next of kin or legal guardian give agreement to any non-emergency operation that is to be performed on a legal juvenile.

If it is your child who is going to see a surgeon, and if treatment is in question, it would save everyone's time if you came with your child, or at the very least your child should be accompanied by a responsible adult, preferably a relation.

If you do not do this, and if treatment is in question, it might be necessary for the surgeon to postpone the whole affair until you do come with your child, and he is able to discuss his recommended treatment with you.

The emotional and intellectual maturity of the legal juvenile is not the question. It is purely a matter of meeting legal obligations related to the juvenile's date of birth, exactly the same issue as obtaining a driving licence or any other legal issue decided by date of birth.

Conduct of the physical examination

After the interview is complete, or sometimes the interview is still progressing, the surgeon will want to make an examination of your body, termed a physical examination. Patients of the sex opposite to the doctor's, or patients of the same sex who are required to strip substantially, are customarily given a gown to wear, or a sheet to lie under. It is considered proper practice for the surgeon of a sex opposite to the patient's to leave the room while the patient changes clothing.

Your surgeon will concentrate his attention on the area of your body that is the cause of your concern. For instance, if you have pain in the upper abdomen, that's the area he will look at, and examine by touching, inch by inch, carefully and in detail. He will, however, examine the rest of the abdomen, and almost certainly your chest as well. If you have a problem with your right knee, don't be surprised that your surgeon first examines the left knee. He wants to find out what the painless knee is like before he makes any decisions about the one that hurts. He very likely will look at your hands and other joints to check if there is any evidence of a generalised joint (arthritic) problem.

Once the examination is finished, the surgeon may discuss his conclusion with you, before you resume your clothes. If it is more complicated, he will suggest that you dress first, and then discuss his plan with him. We are all aware that a patient is psychologically disadvantaged when he is wearing nothing but his socks and a flimsy gown!

Diagnosis

The preliminary to all treatment is first deciding what is wrong, that is, making a diagnosis.

The full diagnosis may be perfectly obvious to all concerned, you and the surgeon, and may in fact have been established before you came to his office. In such a case the discussion usually concerns only how to treat your problem.

But let us suppose that the diagnosis is not fully established, or that although the general diagnosis is known, further information about the problem is required for the surgeon to be able to suggest to you what he would consider the best way to deal with it. He may suggest to you that further tests are in order, or that consultation with specialists in other fields of medicine would assist in making the diagnosis, or in ensuring that you are optimally fit for the surgery, that the risk present in every operation and with every anaesthetic has been reduced to its minimum.

Tests the surgeon might order

In reaching a conclusion about your problem, or your "complaint," the surgeon will in general have one diagnosis that he considers the most likely, and then several others which vary from reasonably likely alternatives to highly improbable possibilities, but they are still possibilities.

Where time permits he will do all he can to shorten this list, known as a "differential diagnosis." Shortening the list may require further tests to be made. It may require tests that were previously performed to be made all over again, since a chemical or structural alteration that did not show on the first tests, the ones you had made a few months ago, might now be found. Or there might be a more refined way of doing the same test with a more complicated (read more expensive) piece of equipment.

Ruling-out tests

Most tests, tests of all kinds, are not ordered because the doctor expects a "positive" result, but because he has to prove that the possible diagnoses, the ones low on his differential list, cannot be correct. These are known as the "ruling-out" tests. If, for instance you are fifty five years old and have recently developed severe and persistent back pain, your surgeon's first and most likely diagnosis on his list is "osteoarthritis of the spinal joints." But he has to prove to himself, and may explain to you his reason, that you don't in fact have a fracture due to osteoporosis, nor do you have a bone tumour in your spine. So he orders a bone scan and some blood tests. These are "ruling-out" tests. They are expected to come back "negative." Of course, the surgeon wants them to be negative, he expects them to be negative, and all they do is rule out the possibility of the already less likely diagnoses, rendering the original "most likely diagnosis" even more probable.

It is important that you, the patient, should understand the "ruling out" nature of the test, or you may express disappointment when it comes back "negative." It isn't kind to tell you that if it was positive you would have cancer, so you really should be pleased when it's negative!

Your surgeon may wish to order further tests to reinforce his conclusion about the diagnosis. These are "diagnostic" tests, and fall into the broad groups of "invasive" and "non-invasive," which means that in the invasive tests someone puts something into you, commonly by a needle, and in the non-invasive tests nothing is put into you, or at least it doesn't go through your skin.

Invasive tests

a) Blood tests

To take blood for testing requires that a needle is put into your vein, and that's thought of as invasive, but usually nothing will be injected into the vein, it's just your blood that's taken out. No one likes to be stuck with a needle, it certainly isn't painless; the best I can say about it is that, even with needles, technology has improved them dramatically. They no longer re-use the blunt needles with hooked ends we experienced in my younger days, and the whole process, although it's never pleasant, is far less unpleasant than it used to be.

b) Nuclear medicine

Nuclear medicine scans generally involve a needle put into the vein, some radio-active substance is injected, and at some time later in the same day when this substance has spread through the body, a picture rather like a miniature X-ray is made. If an abnormal concentration of the injected material is taken up at a particular point, as it might be in a bone fracture for instance, this shows as an abnormal area on the picture, and is termed a "hot spot." Some patients fear anything called "nuclear" but they can be reassured there is no more danger in this test than in any other that is ordered for them.

c) Imaging

What was until recently called the X-ray department in any hospital is now usually called the Imaging department because although they make pictures (images) of your body as they have always done, not all of these pictures are now made with the use of X-rays. Of course, the traditional X-ray studies that every one is now familiar with are still made there.

A CAT scan is a specialised X-ray study. It doesn't use a cat, although any number of cartoons have suggested that. The letters stand for "computerised axial tomography," which means that pictures can be made through the body in "slices" making use of a special X-ray machine associated with the use of a computer. The computer is used to juggle the black and white in the film, and with this technique, it is possible to make a picture of a solid organ, such as the liver, which could not be shown on an ordinary X-ray picture. It is also possible to make a transverse picture of a disc in the spine, to show whether it has been ruptured, to show the nerve that runs behind the disc, and whether it is compressed by the rupture. Most hospitals of any size now have the special equipment needed to make a study with a CAT scan. The use of the computer in this instance is the forerunner of the future in "imaging." It is thought this will likely be entirely *digital* and if you don't understand the meaning of "digital" then ask your child to explain it, he's been brought up in a digital world.

But there are other "imaging" studies where a "picture" of some kind is made, and these also require that you go to the Imaging department.

A picture can be made by the use of ultrasound. Many persons are now familiar with this to show the baby while still in the womb. The technician who makes this picture has usually had her initial training as an X-ray technician, generally now called a radiographer, and has had special extra training in using ultrasound to make pictures. Your surgeon may of course use this technique for the purposes of demonstrating the baby, if that is the issue, but he can also order a study of other abdominal organs, such as the liver, to find out whether it is of normal size, and whether it has any abnormal masses in it, such as cancer or cysts. Ultrasound is not helpful with bone conditions, but has been used to show abnormalities of tendons or ligaments at the joint, and is used to show a dislocation of the hip in a baby at the age of a few weeks, before the bones at the hip joint contain calcium, and therefore before they can be imaged on an X-ray.

The cardiologists use ultrasound to image and follow the function of the heart and its containing sac, the pericardium.

MRI, a type of imaging equipment, is unfortunately not as available to the public in Canada as most of us would wish. The letters MRI represent "magnetic resonance imaging" and the technique involved does not use X-rays. A picture is derived by putting the patient inside what looks like a long tube, which is in effect a circular magnet. The test is entirely harmless as far as anyone knows now, but some claustrophobic patients don't like being put inside this tube, and even the bravest of us find it a little intimidating. The image derived from an MRI is the clearest way we presently have available to us for picturing "soft" tissues, that is anything other than bone, for instance the brain, liver, nerves, spinal cord, knee ligaments and so on. CAT scans are often thought to give better pictures of bone, and MRI scans are better for non-bone studies. Regrettably due to the insufficient supply of MRI machines in Canada, due to limits imposed by government budgets, you and your surgeon will often have to make use of the second best, the CAT scan.

d) Contrast studies

There is a type of imaging study which is known as "contrast." A substance is introduced into a part of your body, which will show up on the X-ray, filling or outlining that portion of the body into which it was injected, if an "invasive" study, or the organ where the material is concentrated as a part of that organ's normal function, if the substance is taken by mouth.

Direct injection into a part of the body, to make a picture of a particular structure, may for example be into the arteries (arteriogram or also known as an angiogram), the veins (venogram), the joints (arthrogram), the spinal canal (myelogram) or the intervertebral discs (discogram).

Pictures can be made by swallowing material which will show white on the X-ray, outlining the gullet (oesophagogram), or can be followed, as it moves through the body, on an X-ray screen by the X-ray doctor as it passes into the stomach and upper portion of the intestine (upper gastro-intestinal series). The same material can be injected via the anus, to outline the rectum, and colon (lower gastro-intestinal series).

The normal gallbladder is outlined when the patient swallows a pill which contains material that is absorbed from the intestine, concentrated in the liver, and passes into the gall bladder (cholecystogram).

The kidney is outlined by injecting material into your veins, this material is concentrated in the kidney (intra-venous pyelogram, IVP) and then excreted through the ureters and bladder.

The tubes of the lung can be demonstrated by squirting material into the windpipe; this is breathed into the lung by the patient, making a very pretty picture, like seaweed waving slowly under water (bronchogram).

Artery studies

Normally it is possible for the doctor to find a pulse in your arm, leg or neck. The pulse represents the flow of blood, pumped from the heart; the pulse rate, the number of beats per minute, is the rate at which the heart is contracting. If the doctor finds a normal pulse, one that can easily be felt, he is reasonably confident that you have a normally functioning artery.

But what if he can't find a pulse at either ankle? Or he can find the pulse at the left ankle but not the right?

It might be that your pain is due to insufficient blood being brought into the limb because of arterial blockage, in that case some direct treatment on the artery might be indicated. Or it might be that you were unaware of any problem with the artery, but the surgeon has found one and thinks it wouldn't be safe to use a tourniquet on your legs, to operate for instance on your knee, until he knows more about the flow of blood in the arteries.

Among the simpler "non-invasive" tests, is a Doppler procedure, where the noise made by the flow of blood through the artery is greatly magnified, and compared against the normal. If the flow is not normal, not only can this abnormality be shown, but the place in the artery where the flow is obstructed can be found.

e) Specific purpose electrical tests

Tests may be performed with electronic devices to measure function in various parts of the nervous system, the brain, spinal cord and peripheral nerves. The commonest of these are nerve conduction and electromyogram (EMG) studies to examine function in the limbs, electro-encephalography (EEG) to examine brain function, and evoked potentials studies to examine function in the spinal cord and cranial nerves. The cardiogram (ECG or EKG) is too well known to need description, but there are shorter and longer cardiogram assessments, according to need, commonly classified as 2 lead (ultrasimple), 6 lead (the usual), 12 lead (in special circumstances).

General health tests

If there is any question that you might need an anaesthetic, other specific tests, related to your general health, the function of your heart and lungs, might be ordered. An anaesthetic, and an operation are all stressful, you will need the best possible function in your heart and lungs. The surgeon, or other doctors working with him may order tests to see how well your heart and lungs are working.

The commonest of these, to the point that it is routinely ordered in patients of middle and older years, is the electrocardiogram, usually abbreviated to ECG, or sometimes following older German terminology, EKG. It is really a tracing of the electrical impulse passing around the heart muscle as it stimulates it to contract. The ECG gives a great deal of information of what *has happened* to the heart in the past, but a completely normal ECG *does not mean* there's nothing wrong with your heart at this moment, or that there won't be in the next few minutes. It's a useful test, but it is not an infallible indicator of the future.

Pulmonary function tests are less commonly ordered. They may be performed in a private office, but are more commonly made in a special laboratory in the hospital, and most commonly are ordered by the anaesthetist or a physician who specialises in lung diseases, a "respirologist" or "lung doctor." These tests may be ordered if an operation is under consideration. They involve testing the volume of the flow of air in and out of the lungs, the ability of the lungs to perform their normal function to absorb oxygen, and to rid the body of waste gases, all problems most commonly faced by patients who have

51

spent their lives wrecking their lungs with tobacco. Smokers aren't the only persons who get lung troubles.

Other consultants

Your surgeon may wish to obtain the advice of other specialists, and this will involve further appointments, and it is regretted, further delay in coming to a final decision on what is wrong with you, and what should be done about it.

These further consultations with specialists are of two kinds, they might be:

- Diagnostic or
- Pre-anaesthetic assessments.

If diagnostic, that is to say, the surgeon thinks other specialists in the same field as himself, or in other areas of medical specialisation, should be consulted, then these arrangements will often be made by your family physician, following the surgical consultant's recommendations.

If pre-anaesthetic assessment type consultations are indicated, these will usually be arranged through the surgeon's office, with for example, an anaesthetic specialist (anaesthetist), a lung specialist (respirologist) or heart specialist (cardiologist).

Pre-operative improvements

As part of the plan of treatment, whether it does or does not comprise an operation, the surgeon may recommend to your family physician that you undergo a programme of treatment to improve your lung function (respiratory therapy), your heart function (cardiac therapy), your muscle and joint function (physiotherapy), and weight alteration (usually to make it less, occasionally more!).

These programmes will usually be arranged by your family physician, with one or more different types of therapists.

Almost certainly you will also be advised that you should, for health reasons, give up smoking. Forty years ago I was about the only doctor I knew who didn't smoke; nearly all doctors have now given up smoking, and are very keen that everyone else should do the same!

DECISION TIME: WHEN ALL TESTING HAS BEEN EXHAUSTED

There comes a time in this process when a decision has to be reached in regard to giving your problem a name, a *diagnosis* and deciding what to do about it.

The diagnosis may lead the surgeon to recommend that you be treated by means other than an operation, variously called non-operative, non-surgical or conservative treatment.

I have often known patients to misunderstand a decision of this kind. Whereas it was the surgeon's statement to the family physician that treatment could be given quite adequately without any operation, the patient converts this statement to, "The surgeon said he *couldn't* operate."

It is quite possible that even before you saw the surgeon, the diagnosis may have been perfectly obvious, for instance a hernia (rupture), or varicose veins, and all that is required is a discussion with the surgeon about the operative treatment you fully wish to have, and came to him to arrange.

Or the diagnosis may now, as a result of the tests that have been performed, have been established, the diagnosis has to be explained to you, and the type of treatment needs to be discussed.

Or it may only be obvious in which part of the body the problem lies, but even after all the tests we've described have been performed, it is still not possible for the surgeon to be absolutely certain of the cause of the pain. We have many tests and forms of testing, but contrary to common belief, we do not have a test for all conditions, and the surgeon even after ordering all the tests known to the profession, may still be in the position of saying *I think this diagnosis is the most likely, but on the other hand it is possible you have this or that or the other*. For instance he may tell a woman of fifty years that he is fairly sure her right sided lower abdominal pain is due to acute appendicitis (an infection of the appendix) but that it might possibly be due to a problem with a cyst of the ovary, which lies right next to the appendix and results in pain and tenderness in the same area, or it might even be due to a cancer of the caecum (a part of the large intestine located in the same area of the lower right abdomen).

Or it may not be at all obvious where the problem lies in the body. None of the tests have been conclusive or even remotely helpful except in "ruling out" other possible diagnoses, and the surgeon may have to resort to suggesting an exploration of the whole of the painful area, known in the abdomen as an "exploratory laparotomy," or in more simple English, making an incision and examining the contents organ by organ, and the length of the bowel, inch by inch, with reasonable expectation that the cause of the problem will be found, but possibly even at this exploration it may be established that there is no physical abnormality that can be found to explain the pain.

What to do?

Deciding *whether* to treat the symptom (complaint) is the next decision you and your surgeon will make together, he will spell out as best he can, what may happen if you don't have treatment, and what treatment he would recommend if you do wish to be treated.

Whether you do wish to have treatment is *your* decision, treatment is always optional for any condition, it is *never obligatory*, be it a cancer or a fracture. As long as you are considered legally competent, by age and mental status, ***it is your decision whether you have any treatment.***

Treatment as determined between you and the surgeon will be either:

- Non-operative or
- Operative

Non-operative treatment will not involve an operation. Depending on the surgeon's specialty, few or most of his patients are treated without surgery. A general surgeon is more likely than some other surgical specialists, to deal with persons who have visible abnormalities such as varicose veins, piles or a hernia, or who have been fully investigated by medical specialists such as gastro-enterologists (internists who specialise in problems of the stomach and intestine); such patients come to the general surgeon with their minds pretty well made up that they want to have an operation. In consequence a high proportion of general surgery consultations are preliminary to an operation.

In contrast, the orthopaedic surgeon is often viewed as an appropriate person to make a diagnosis of a limb or back pain, to prescribe treatment for it, with the expectation that a very much smaller proportion of his patients will be operated on. This is the basis for separating non-operative from operative treatment, and as emphasised elsewhere, agreeing to see a surgeon is not the same as agreeing to have an operation. You see the surgeon to get a diagnosis, and to learn what the options are in regard to dealing with this diagnosis, that is to say, how your problem can best be treated.

We are concerned in this book with operative treatment, and will put aside the non-operative treatment. This will usually, acting on the advice of the surgeon consultant, be carried out under the direction of your family physician.

I cannot emphasise enough that non-operative treatment is the bulk of many surgeons' practices.

It is therefore reasonable at this point to ask the question that we've been walking all around...

WHAT EXACTLY IS A SURGEON? WHAT IS SURGERY?

An alternative word for surgery, one that was often used in rural areas, was "cutting." Surgery generally implied a cut, and generally implied that some part of the body was cut out, or removed. Within my own experience of work in rural and outport coastal Newfoundland, forty years ago, I recall hearing doctors spoken of as a "very good doctor, does lots of cutting."

Although not now an adequate definition of the role of the surgeon, the original use of the word "surgery" defined it as the treatment of disease or injury, by the hand.

THE "HOW" OF SURGERY

Open, semi-open and closed surgery

These terms are not self-descriptive, they are in widespread use, and are the best the profession has managed to come up with so far. You are welcome to try your hand at inventing better ones for us.

"Open" surgery is the traditional kind of operation, the type that most people will think of. It involves making a cut into the body sufficiently large that the edges of the cut can be pulled apart (known to the surgeon's assistant, who gets to do this part of the job, as "retracted") and the underlying organs of the body exposed. Various names are given to the procedures, usually a combination of the area of the body as the first syllable e.g. thorax, and "otomy" for the second syllable, indicating an opening was made into it. Examples are thoracotomy for the chest, craniotomy for the head, and arthrotomy for the joints.

The advantages of an open operation are that the area of concern is quickly demonstrated, it requires the least amount of expensive equipment, and the surgeon isn't going to find in the middle of his operation that sophisticated electronic technology has failed him.

The disadvantage of the traditional "open" operation is that it requires a large incision to be made through healthy tissues to gain access to the unhealthy underlying tissues, that the wound created is subsequently painful, and that this incision (for which read "wound") commonly results in the need to spend several days in hospital, and several weeks of convalescence, before the patient is able to resume the activities he was pursuing before his operation.

The surgical profession went through a period of "keyhole" surgery, operations performed through very small incisions. Those who practised this form of incision were able to assure their, usually "society," patients that the scar would be small; those who didn't practise it used the slogan, "Big incision, big surgeon," the obvious implied opposite of this remaining unspoken.

Endoscopic surgery

A system of surgical treatments, using what is effectively a telescope, has been developed to examine and treat many of the body's abnormal structures; a very small incisions (about 1 cm) is adequate for the insertion of the telescope, and one or more incisions for insertion of instruments, thus reducing the size of the wound, the damage to normal structures on the way into the body, and the period of disability after the surgery.

For many years we had illuminated rigid tubes, with little or no magnification, which were used to look down into the stomach, or the lungs, or up into the rectum, and were no pleasure for patient or surgeon. Technological improvements with fibreglass optics allowed the development of flexible instruments, with telescopic lens systems, which could be inserted with much less discomfort, and much further down or up the stomach and intestine. These are at present most commonly employed for diagnostic purposes, but surgery of a relatively restricted character can be performed with such an instrument, inside the cavity of the stomach or intestine, a technique called *endoscopic surgery*. The 'endo' means it's done inside; the 'scopy' means an instrument like a telescope is used.

Claims of who did anything first are always hard to establish, in particular when national pride interferes with scientific judgement and memory. However, it was the specialist bladder and prostate surgeons who first brought this technique of operative endoscopy into their standard therapeutic surgical practice, saving the patient, who was often old and infirm, from having the bigger and bloodier operation of suprapubic prostatectomy still performed by the surgeon who was not a specialist in this field. Their instrument, a telescope inserted through the penis canal (urethra) and up into the bladder, and known as a *cystoscope*, was the forerunner of our now widely practised endoscopic surgery.

There were occasional examinations of the knee joint made with the use of a child's size cystoscope, but the original purpose-designed instrument came from Dr. Watanabe in Japan, brought to North America and Europe by a Canadian surgeon, Dr. Robert Jackson. It revolutionised knee surgery; whereas prior to endoscopic surgery (known in a joint as *arthroscopic*) the patient had an incision a few inches long, spent a few days in hospital, followed by physiotherapy for a few weeks while he was off work for the same length of time. With current methods, office workers are often back to work the day after arthroscopy, and star hockey players the following week.

Endoscopy is now employed in other joints, the spine, the abdomen and the chest; it is in some special circumstances used on the head.

Endoscopy is now the commonest technique for operating on the gall bladder, it allows the surgeon to perform the removal of the gallbladder, but through small incisions instead of the previously required twelve inch incision. A similar technique is employed in repair of a hernia (rupture) at the groin, where using a telescope through a small incision, the surgeon can plug the abnormal hole. The advantage is not just a smaller incision, the real advantage lies in less pain after the surgery so that the patient leaves hospital the same or next day.

Surgeons are developing endoscopic techniques for use in other parts of the body, such as the heart, lungs, uterus, and spine. These are still in the developing phase and not all in wide use, nor have they completely replaced surgical techniques with larger incisions.

The patient benefits enormously in having reduced problems after surgery, and reduced time away from his work. The losers, we regret, are our nurses, for whom there is reduced need for post-operative care in hospital, and hence reduced employment opportunities: improved technology always seems to take away someone's job The Luddites had thoughts on this!

Microsurgery

Microsurgery, as defined by surgical usage, is surgery performed with the aid of a microscope. Although patients will often refer to their endoscopic surgery as "microsurgery" this is not the way surgeons employ that term.

Among the first techniques was the use by the surgeon of magnifying spectacles of various kinds. This permitted suturing and repair of small blood vessels and nerves that had been damaged. The next phase was the development of an operating microscope, a large, fixed instrument held over the patient, which permitted an excellent view of the tiny structures that needed to be repaired or replaced, such as the miniature bones in the ear, or nerves and blood vessels in a finger that had been cut off, and was being sewn back on again.

Laser surgery

The laser is a tool, and as such "laser surgery" is not a specialised procedure in most instances. It does offer a potential for treatment in the eye, not possible by other means. The principle advantage of the laser in areas such as the knee joint is the bloodless nature of its use, so that a patient who is a "bleeder" has fewer post-operative problems.

"Blind nailing"

This term "blind" is a misnomer, but describes a form of orthopaedic operation conducted with the repeated use of an X-ray machine in the operating room, which permits a nail to be put down the hollow of a fractured bone, using a very small incision at the end of the bone, and not exposing the fracture itself.

Angioplasty

Operations are performed by the radiologist (X-ray doctor) on the interior of blood vessels, including those that go to the heart. This represents an area of what is, effectively surgery but it is performed by persons who are not designated as surgeons. The distinction, the boundaries, between the fields of specialisation is becoming blurred.

Biopsy

If your surgeon is uncertain whether a swelling is or is not due to cancer, he may wish to remove a small part of this to have it examined under the microscope by a pathologist; the removal of only a small part of the tissue for this purpose is called a biopsy. Depending on its location and other factors, the biopsy may be made through a small incision, or by means of a hollow needle.

THE "WHO" OF SURGERY

The team

Surgery, the "cutting" part of the operation, is performed by surgeons, but the surgeon will not be working alone. He needs a team to support him. Some of the team members are seen to be present in the operating room, others, just as on the stage, are essential "behind the scenes" players.

The surgeon

Although any medical doctor is in theory licenced to perform any surgical operation, in most hospitals in North America now, the person performing the surgery will have had special training and taken examinations to show that he is particularly qualified to perform surgical operations. Training to be a surgeon generally takes five or more years of study after completion of the MD degree; even after that there may be another year or two pursuing studies in more narrowly specialised areas. Your surgeon therefore is likely to have studied his "trade" for at least twelve years after graduating from high school, before he enters independent practice.

In Canada, the surgeon takes examinations conducted by the Royal College of Surgeons of Canada. These examinations differ in content according to the surgeon's area of specialty, that is, a brain surgeon writes different examinations from a bone surgeon. After the required five years of post-MD training and successful completion of his exams, the surgeon is designated as "certified" in his area of surgical specialty, e.g. gynaecology, general surgery, neurosurgery, but all surgeons use the same initials after their names, FRCS(C), meaning Fellow of the Royal College of Surgeons of Canada

In the United States there is a similar programme of training; in fact the surgeons' organisations of the two countries have co-operated to arrange a training system standardised between Canada and the USA. In the USA, each surgical specialty has its own Board of Examiners. After completion of the Board examination the surgeon is designated as Board Certified, but, unlike his Canadian counterpart, he does not have any additional initials to put after his name to indicate this. Thus, whereas a Canadian surgeon is usually designated MD, FRCS(C), an equally and fully qualified American surgeon displays only the MD initials.

Anaesthetist or anesthesiologist
Not many of us would wish to have surgery without some method of relieving the pain that it causes. The pain may be prevented by the surgeon injecting a "freezing" or "local" anaesthetic around the nerve

that supplies sensation to the area of the operation, this is what your dentist does when he operates on your tooth.

Most, but not all, surgical operations require the services of an anaesthetist. The anaesthetist may use a number of different ways of relieving your pain, these will be discussed later. He is also the person who supervises your general state of health during the operation, watching the cardiogram, the pulse rate, your blood pressure and so on.

We don't any longer have a system in the operating room of "Captain of the Ship." Each member of the team makes his own decision, and each member respects the decision of the other member. As a result, no operation is started by the surgeon, until the anaesthetist believes the patient is suitably fit, and the surgeon respects the anaesthetist's advice throughout the operation.

In areas using British terms, such as the United Kingdom, Ireland, Canada and Australasia, the anaesthetist is a medically qualified doctor, that is, an MD, and he has taken approximately the same number of years to become a specialist as has the surgeon. In the US, as in Sweden, France, Belgium, and a number of other European countries, some nurses are specially trained for this function. We therefore find in the US that nurses who give anaesthetics are called anesthetists (in the US the "a" is dropped), and the board certified MD specialist who supervises their work or gives the more complicated anaesthetics, is termed an anesthesiologist. Effectively, the Canadian MD anaesthetist specialists, and the American MD anesthesiologists, have had the same training, given over the same number of years.

In smaller hospitals in Canada many or all of the anaesthetics are administered by family physicians who serve a dual role as both family physician and anaesthetist in their community. They have not spent the same number of years in training in giving anaesthetics as a specialist will have done, but they provide a service essential to the medical functioning of their community.

Surgeon's assistant

If you think of the very smallest surgical procedure, for example stitching a one inch cut, you will understand that the surgeon can do this without any help, except the nurse who knows where the equipment is stored, fetches it and lays it out for the surgeon (poor helpless fellow!).

If you think of the very largest of operations, for instance a heart transplant, you will understand that a surgeon often needs more help than just a single person to help him lay out his tools.

Other operations lie somewhere between these two extremes, and the need for assistance in an operation, like everything else, has a supply and demand basis. In a university hospital where teaching is a major component, in the performance of your operation the surgeon is likely to have a trainee at the operating table, whether he has the need for him or not. In remote areas the surgeon might very much prefer to have some skilled help, but it simply isn't available.

Within my own practice experience, it was expected that the patient's family doctor would make house calls, deliver babies and assist at his own patients' operations. That is still the norm in rural practice, but is rarely found now in urban practice. In non-university urban hospitals, the surgeon when he has the need is likely to be helped by another surgeon (particularly if someone who fully understands the operation is necessary, or if he is semi-retired) or by one of a handful of family physicians to be found in each community who enjoy being in the operating room and are pleased to act as assistants on cases that are not their own patients.

In some ways this is an improvement on the old system, where the patient's own doctor would often not come until well after the operation had started, he'd talk all the way through the operation about irrelevant social matters, be indifferent to what was needed of him as an assistant, leave before the operation was finished, and expect to be profusely thanked for having referred the case. There developed from this a surgeon's saying, *stop assisting and start helping!*

Nurses

It is unimaginable that a surgeon would operate without the assistance of a nurse, the truth is he probably couldn't even find the tools he needs to do the job!

As a patient you will encounter nurses at every stage of your progress through the hospital. They are present in the emergency department (now often called ambulatory care) if that has been the basis of your admission. They now do pre-admission interviews and assessments to take your medical details when you come into the hospital, they will look after you on the ward (which will be discussed later) and they are essential to the operation.

Surgical operations have become more and more complex. A surgeon in Nelson and Napoleon's time could, and did, carry all his tools in his handbag. Nowadays any operation we do is likely to need at least one long table covered in instruments. It is the nurse's function to be sure that the surgeon has the instruments and tools that he needs, to get these tools sterilised and ready for the following operation, and to provide some assistance to the surgeon during your operation.

Most nurses who work in the operating room have had specialised training, sometimes in a university hospital course, sometimes community college, and often "on-the-job" training. Since every surgeon has his own way of doing things, no matter where the nurse had her first operating room training, a component of "on the job" will be required in every hospital she goes to.

The relationship between surgeon and nurse is never quite as intimate or quite as much rollicking fun as portrayed in MASH, but it does require a considerable degree of mutual understanding, in order that the nurse can know what the surgeon wants and can have it ready for him. There is an old operating room joke, "Don't give me what I ask for, give me what I need!"

Technicians

Except in the very smallest of hospitals, the operating room nurse works only in the operating room. She does not do "bedside nursing." Whereas the bedside nurse knows little or nothing about the instruments and tools the surgeon uses in the operating room, the operating room nurse is rarely if ever called on to employ the skills of the bedside nurse. In the same way as the anaesthetist and surgeon serve different but essential roles during the operation, operating room nurses and bedside nurses simply have different functions.

To the cost conscious hospital director, a person with limited training in operating room work can be employed less expensively than a fully trained nurse who has post-graduate training in OR skills. These persons, commonly called OR technicians, can do most of the work of the OR nurse, with the exception of administering medication, and taking charge of the unit. The average larger OR suite has now a mix of nurses and technicians.

As in every other walk in life, how well the individual does his work is a measure more of the individual than the title he bears or the pay he takes home. Most surgeons are unaware of who is the nurse and who is the technician.

Secretaries, medical records and lab techs

These less recognised categories of "behind the scenes" hospital workers are all essential to the smooth running of any operating room, and all are essential to the "beginning, during and after" phases of your operation.

Emergency Surgery

This is the kind of surgery you were given very little choice about. It can be either:

life saving,

or limb saving,

or baby saving.

Life Saving Surgery

This is the kind of operation that is needed under the most severe of emergency conditions. The patient may or may not be conscious. If conscious, he can refuse the operation and choose to die, that's his business. If unconscious, and if there are no responsible persons available to make the decision, the surgeon and his team have a personal, professional, ethical, as well as a legal responsibility to do all they can to keep the patient alive. Most, but not all, of the really urgent but unconscious causes are related to bleeding.

Examples are:

Severe head injury with progressive bleeding compressing the brain.

Ruptured uterus in obstructed childbirth.

Ruptured and bleeding spleen or liver from abdominal road accident trauma.

Limb saving surgery

Whereas most life saving operations are due to blood flowing out of the vessels and going where it doesn't belong, most limb saving operations are performed because the flow of blood in the limb is obstructed.

In non-accident cases, we find usually in older persons that a clot or the lining of an artery may detach and the flow of blood to the limb is cut off. If this is not cleared within a few hours, part or all of that limb may die, requiring partial or complete amputation.

In accident cases, the artery might be obstructed or torn, either as a direct injury to the artery itself, or as an injury to a bone or joint which then compresses or tears the artery. Early operation is essential to preserve the limb.

There are infections that cannot be controlled by the use of antibiotics alone. Pus may build up and may need to be released from the body by making an incision to let it out. This is perhaps now most commonly seen with infected joints, and in less wealthy countries is still very common with infected bones. Although it gets a lot of attention in the press when it occurs, the horribly named, and horrible in fact, condition of "flesh eating disease" is fortunately very uncommon, in this condition a type of spreading infection advances at visible speed up the limb, it cannot be controlled by antibiotics. If the limb is not cut off (amputated), this infection will overwhelm the whole body, resulting in death within a few hours.

Baby saving surgery

During a pregnancy, proper care of the child requires periodic monitoring of the rate of the baby's heart beat, of the baby's movements and other activities. As the delivery (labour) begins, the child is at greater risk because, among other things, the blood supply it receives, which carries the oxygen that is essential to its survival, depends on the after birth (placenta). If the placenta begins to separate prematurely, or one of a number of other things that can go wrong at this time, the baby's heart rate begins to falter, movements decrease and other signs of "foetal distress" are encountered. It is then a matter of judgement, employing several well known and well accepted guidelines, whether an emergency operation should be performed to save the baby's life, or to save the baby from permanent brain damage. This operation usually goes by the name of a Caesarean section, since the Roman emperor, Julius Caesar, was said to have entered the world in that fashion.

Urgent surgery

Urgent operations are the ones you and your surgeon have time to think about, to discuss all you please, but can't usually wait for your summer holiday to have them. They generally are given a degree of priority in the hospital, but are not the "bump everything else on the list" priority life-saving operations like an emergency Caesarean section. They are performed for malignant conditions (cancer)

non-malignant conditions.

Surgery for malignant conditions

It is frightening for anyone, no matter how knowledgeable, no matter how well controlled, to be told, "You've got cancer."

The natural first reaction is "Well, cut it out, operate now!" There was a time when cancer operations were performed almost as emergencies. We now know, very much to our regret, that once a cancer has been diagnosed it is almost certain that at least some of the cancer cells have spread through the body, and there is no question of it being a rush against time to cut the tumour out.

In consequence, your surgeon will take trouble and spend some of yours and his time to make sure that your operation is going to be done in the best possible manner, under the best possible conditions, with the best possible team available, and with the best possible combination of all the treatment methods of surgery, chemotherapy, radiotherapy, nuclear medicine and anything else that will benefit you.

For all of these reasons, surgery for cancer is usually done "soon" rather than "now."

Surgery for non-malignant conditions

Life can be threatened by more problems than cancer, and the diagnosis creates the same high level of anxiety.

Your surgeon may tell you that an operation on your heart is essential, and that you can expect to die suddenly if you don't have it. There is, of course, an urgency to say, "Do it today, do it now, don't wait till I'm dead!"

The surgeon is unlikely to agree.

A major factor restricting him is the need to assemble his team, in this case a very large and very highly specialised team, with an unusual amount of equipment. His operating room is severely restricted in how many operations it can undertake each week.

Another restricting factor is the need to see that you are in the best possible health before the surgery is performed. Often a person with heart disease will have other problems such as diabetes, and all of these need to be brought under the best control possible to give you the very best results in your operation,

Elective surgery

This is the kind of surgery you can reasonably decide whether you wish to have or not wish to have, that is, neither life nor limb are threatened. The problem may be causing discomfort which you are or are not willing to put up with. It may be limiting your ability to get around or work, and you may or may not be willing to put up with that.

The surgery cannot be offered to you before the surgeon is able to arrange it in the hospital. It is not an emergency or urgent operation, and it will be fitted in when the surgeon is offered time by the operating schedule. This will give you the opportunity to arrange with the surgeon for a hospital booking that is least inconvenient to you. You probably won't be able to say, "Friday the third of April at two o'clock," but you will be able to say, "Not April, I'm getting married that month!"

Cosmetic surgery

Variously known as cosmetic or aesthetic surgery, this is a type of operation that is not concerned with the saving of life or limb, it is not concerned with the improvement of function or the relief of physical pain. It is devoted to the change of outward appearance. Most commonly these operations are performed by specialists in plastic surgery, who will also perform in their practices other operations that are very much concerned with the improvement of function.

Distinctions between the different purposes behind cosmetic surgery is more related to what an insurance schedule will or will not pay for, and the distinctions sometimes get twisted around so they will meet a definition in an insurance schedule.

One type of surgery is to repair a deformity. There are a number of conditions which an unfortunate person may be born with, or which may develop in later life, and which although they do not threaten life they may seriously interfere with an individual's willingness to be seen in public. The book, later the film, *Elephant Man*, made this widely known.

Another type of cosmetic surgery is to repair an injury which does not interfere with function. A person may have suffered a burn to the face. The face is distorted by scar, but there is no interference with function.

SURGERY FOR IMPROVING ON NATURE

The type of cosmetic surgery that most persons are familiar with is the variant known as "aesthetic." This is largely to correct the ravages of time, for instance to raise drooping facial skin, or bags under the eyes. To a lesser extent, to correct an appearance that some persons might find perfectly acceptable, for instance the shape of one's nose, but the owner of the nose would prefer it to be different.

MAJOR AND MINOR SURGERY

Patients often ask, "Is this major surgery?" Or hopefully state, "It's only a minor operation, isn't it?"

There are some black humour answers in this area, among them is "minor surgery is done by minor surgeons." More relevant to you, is "minor surgery's what someone else has!"

There is no such thing as minor surgery, if the term "minor" is intended to give the impression of "risk free." The inconvenience might be minor, less time in hospital, less time off work afterwards. The pain might be less in an operation to repair a hernia than in an operation to excise part of the bowel. The statistical likelihood of complications might be less in having a hernia repaired than in having a heart operation. However, the totally unexpected does from time to time happen, so no patient should ever be allowed to believe that a "minor" operation is a completely "risk free" operation. If a surgeon tells you your operation is "risk-free" and you find he really intends you to believe that, you might consider looking elsewhere! Just as there are no free lunches, there are no totally "risk-free" medical procedures or operations.

YOUR UNDERSTANDING WITH THE SURGEON

Moral contract

This is not a book on legal advice, and the reader should not assume from these statements that legal advice is being given. However, when a patient and a surgeon agree on a course of action this becomes at least a kind of ethical agreement if not a legal one. As in all agreements, it is proper for both parties to the agreement to understand what the other person believes that agreement to mean.

The agreement between you and your surgeon is not going to be written as a contract. There is not going to be any written document whose words can be argued over afterwards nor will there be any fine print that was never read. But you as the patient are embarking on perhaps one of the most important actions of your life, and you ought to have a clear understanding of what is involved.

Your expectations

What are you expecting this operation to do for you? You should be quite clear about this in your own mind, and your expectations should be clearly understood by your surgeon. A lack of understanding of the limitations of surgery is often the forerunner of a patient's disappointment, or sometimes the disappointment of the patient's relations. The patient knew he might not be able to return to the workforce after an operation, but his son did not.

A stomach operation on the patient with severe ulcer pain is likely to alleviate his pain, but is not likely to permit him to eat as much of whatever he likes in the future; a hip replacement operation is likely to relieve the pain of a severely arthritic hip, but the patient may have to restrict his squash game. The surgeon knows about this, he lives in a world of limited benefits from his skills, he may presume you know about it too. You should ask him to what extent you can reasonably expect to be relieved of the pain or the loss of function for which this operation is proposed. To what degree should you expect to have the same symptoms or problems after the operation?

These are not hostile questions. They do not imply doubt about the surgeon's capabilities. They are simple requests for brief factual answers to assist you in coming to a decision about any suggested operation.

What will the operation be?

Technical terms, often called jargon, are necessary in every trade. If the surgeon was going to use ordinary speech, lay terms, for his operations when he is talking to other members of the nursing and medical professions, he would need to replace a two word term with a two sentence paragraph. But when he books his patient in the operating rom for *"endoscopic cholecystectomy, poss. lap"* does his patient know that this means his gall bladder is going to be removed, that the surgeon hopes he will use only two or three small incisions and will employ the telescope (endoscope) but possibly (poss.) will need to make a full twelve inch incision to explore by open operation through the abdominal wall (lap = laparotomy), and the nurses should have sterilised instruments available and laid out on tables against both of these possibilities.

If the surgeon has not explained the operation he proposes to you, in terms you can understand, you are reasonably entitled to have him do so. No doctor expects you to have a full knowledge of human anatomy, but most persons have enough to know in general terms what it is they are being told.

On the other hand although fortunately few in Canada, there are some patients who believe that they are entitled to a detailed anatomy lesson, complete with coloured slides, models and so on, and a blow by blow description of every step of the operation, and all of its alternatives. By today's standards of political correctness, if the surgeon agrees with this, he is "patient friendly" and if he doesn't agree with it he is "paternalistic." Your own levels of demand for attention and explanation are likely to be known to your family doctor, and this is one of the areas that I have referred to when talking of a "match" between surgeon and patient.

What are the alternatives?
Alternatives to any suggested surgical procedure, are treatment by non-operative methods, for instance by medication or physiotherapy, or treatment by a different surgical operation. I will presume that by this

stage you have already agreed with your surgeon that non-operative treatment will not be adequate.

It is reasonable for you to know, and not particularly time consuming for your surgeon to explain in a few words, why he is suggesting this particular procedure to you, and not another procedure or another method of accomplishing the same end. For instance, why is he suggesting an abdominal hysterectomy when you hear a vaginal hysterectomy doesn't leave a scar? Why an open hernia operation when you hear it can be done with a telescope? If this is an operation for cancer, how much of your body needs to be cut away? All the breast, or only a part of the breast? Are there alternative ways of accomplishing the same end? Would these alternatives be better for you, in the sense of less inconvenience, less time away from work or family? You can be reasonably assured that if your surgeon thought an alternative method would be surgically more effective, then he would be using it; he may, however, have taken into consideration only the clinical benefit of the surgical procedure and not your particular type of work or your family situation.

In what way might I expect to be improved if I do have this operation?

Surgeons are given to knowing that for "condition Y" we all do "operation Z" and suppose that everyone else thinks as we do. They are sometimes a bit surprised when a patient asks, "Why are you going to do this operation? What's the purpose of it?"

You, the patient ought to know the surgeon's expectations. You ought to know whether the operation is expected to get rid of all of your pain, or merely reduce the pain. You ought to know if the cancer will be cured, or its progress merely retarded. You ought to know whether your back will be strong enough to return to your previous labouring job.

Your surgeon can only tell you about the usual results. Since he isn't able to predict the future, he cannot say for sure what will definitely happen to you after the surgery, but he can tell you what happens to the average person on whom he performs this operation. In this way your expectations of what the operation might do for you, and his experience of what the operation has done for other persons, can be matched.

Regrettably it is not uncommon for a patient to have expectations from his surgery that differ totally from what his surgeon knows is possible, a patient who used to play championship tennis is not going to return to this level of competitive play after a knee replacement operation.

What would happen if I don't have an operation?

If something as undesirable as going to hospital, having an anaesthetic, and somebody cutting into your body is proposed, it is only reasonable to know what might happen were you to say, "No. I don't want to have this operation."

There are three possibilities if you don't follow the surgeon's advice, the condition might remain the same, or possibly get better spontaneously without treatment, or it might get worse.

Fortunately for both patients and therapists, time cures many conditions, but let us suppose that you have tried all other non-operative remedies, allowed enough time for nature to take its course, and you find that your problem has not improved with what doctors call *the magic tincture of time*. The question you are now faced with, is whether the condition will get worse if you don't have an operation. You may feel at present that to have an operation would be highly inconvenient, you've just started a new job, your mother has come to visit, all kinds of reasons. If you put the operation off for a month or two, for a year or two, will it be as successful? Or, to get the best results, does it have to be done now?

These are very fair questions to put to your surgeon.

Will I need a blood transfusion?
Blood is given its characteristic red colour by a pigment called haemoglobin, that is carried in the cells. If you separate the cells from the blood, there is a pale yellow fluid left, called serum. The function of the red cells is to combine with oxygen in the lung, then to transport that oxygen around the body, shedding oxygen into the tissue of the body as the haemoglobin is carried in the blood stream, and then

picking up the waste carbon dioxide, to shed it in the lungs. The haemoglobin is therefore the body's transport system for carriage of essential oxygen, without oxygen death occurs in a few minutes.

Blood must also have volume to circulate, if you think of the water in the pipes in your own house, you will understand that there has to be a certain volume of water to give pressure in the pipes to get the water to circulate.

At an operation, despite all the care the surgeon can bring to his work, there may be a loss of blood. Depending on the operation, this loss might be slight or might be so severe that the patient goes into shock, a condition in which the heart can no longer maintain the circulation of blood. The loss of volume can be balanced by giving other fluids; the loss of haemoglobin can not. We have no artificial substance that can do the work of haemoglobin, when we do have one, then blood transfusions will rarely be needed. Until that time your surgeon may advise you that there is a risk of blood loss in the operation proposed, and that it would not be safe to perform the operation unless blood for replacement of the loss was available.

There are conditions which cause a slow chronic loss of blood, such as some cancers, leading to anaemia (a condition in which the haemoglobin is lower than normal). Your surgeon may wish to order a blood transfusion before your operation in order to maximise your opportunities for safe surgery.

Some hospitals have facilities available for patients to give their own blood, to be set aside for the operation in a few weeks time, during which their body replaces the blood taken from it and put in storage. There are many potential advantages to the patient in such a system.

Risks and informed consent

There is an element of risk in getting out of bed in the morning, but there is also some risk in not getting out of bed. There is a risk in driving a car, but there is also a risk in going by bus. We are at risk of something unpleasant happening to us every second of our lives. We can reduce risk by travelling in a safe car and by not travelling in a car without brakes, but we cannot be sure that the other car has good brakes and we cannot altogether eliminate risk.

The surgeon is obliged by law to see that his patient understands that the proposed treatment has in it an element of risk. I don't intend to write about the legal aspect of this, but it is known technically as *informed consent* and means that you are aware in general terms of what you have agreed to have done to you, and what might go wrong if it is done. You will at some point before the operation be asked to sign a form (only one of several forms you will be asked to sign!) to the effect that you understand what the surgeon is going to do and what risks are involved.

Just as you can't cross the road without putting yourself at risk, you can't have an operation without taking a risk. You do know that some roads are more dangerous to cross than others, and you take less chance at a pedestrian crossing than in the middle of a six lane super highway.

In the same way, you do know that with all the precautions the nurses, surgeon and anaesthetist will take, your risk is minimised to the extent possible, but it is never completely abolished, any more than it is when you are crossing the road.

In all activities in life, in engineering, in finance, in buying a house, in getting married, there is a cost/benefit ratio, otherwise called a risk/reward ratio. In surgery this means, "What do I stand to gain; what do I stand to lose?" And, "how likely am I to gain; how likely am I to lose?"

There are, fortunately rare, situations where the patient will surely die if he's not operated on. In such a situation most persons accept any degree of risk.

There is one well known legal case where it was shown that the chances of reward were equal to the chances of loss; and the patient had not been told thay; in most operations the chances of harm from an operation are much less than the chances of benefit, but these chances must be balanced by the informed patient.

Nobody remembers what they've been told
Your surgeon will do his best to answer your questions, but in anxious situations of this kind, when dealing with matters that are outside your

usual experience, and a lot of answers coming to you in a short period of time, you might find it worth your while when you've gone home to write it all down, so you have a better record of what you were told, than trying to rely on your memory.

Guarantees

Your stomach is not an electric kettle that you can take back to the store with your warranty certificate and demand a replacement if it doesn't work. I have never known a surgeon to give a patient a "guarantee" although I've known very many patients who said they wouldn't have the operation *because the surgeon wouldn't guarantee it*. My experience has suggested to me that these were patients who preferred their disabling problem to the possibility of a cure.

My advice on the subject of guarantees is two-fold. Don't expect your surgeon to give you one. Go to someone else if he does.

What will the scar be like?

When I ask a patient if they have any questions they wish to put to me, the most frequent by far is, "What will the scar be like?" I've never been able to decide if the appearance of the scar is uppermost in their mind at the time, or whether we've already covered every other subject of importance.

Appearance of scars obviously matters, and by no means only to women. We know that there are some scars that will be more obvious than others, because of their position in the body; we take trouble to make the scar as little obvious as possible. But *cuts leave scars*.

Sometimes after an operation the scar will get thicker, or may spread sideways. In a small number of cases a second operation is performed to improve the scar's appearance. This is always a possibility but is not often requested by the patient.

How long is the wait for surgery?

Having made, or come close to making, the decision to have a surgical operation, the patient is often surprised to find that it can't be booked on the day of the week that would suit him best, and possibly can't even be booked on any day for another six months.

There are a number of reasons for this, mostly related to Canadian provincial government budgets. This book is not intended to be a diatribe on government incompetence, but you should be aware that a factory such as Ford Motors believes that the greatest efficiency of use of their extremely expensive plant comes from running it round the clock, seven days a week. By contrast, the government budgets are progressively reducing the hours that the extremely expensive equipment in the operating room can be used. In most hospitals the operating room starts work at 8 a.m., stops at 3 p.m., and for only five days a week. That is, the tools and equipment are in use for 35 hours a week, and are unused for 133 hours. Since the hours your surgeon can work in the operating room are reduced, the number of patients in the hospital are reduced, so in consequence are the numbers of nurses needed to treat the patients. Most of us feel that this has also resulted in an *increase* in the number of administrators to deal with the work created by *decreasing* the use of the hospital. *This view is supported by the statistics most recently issued by the government of the Province of Quebec which show that between 1982 and 1996 the rate of increase of costs of administration rose six times faster than the increase in money spent on patient care, and that between 1993 and 1996 there was an actual annual* **decrease** *of 1.5% in spending on* **patient care,** *compared with an annual* **increase** *of 8.7% on* **administration** *costs!*

Whatever the explanation or justification, you are likely to find that although your surgeon would be more than pleased to arrange ("book") your operation when it would best suit you, he is able only to offer to you what the hospital (government allocation) budget makes available to him.

Will my operation definitely be done the day and hour it's booked?
By "booked" we mean, "arranged," every operating room keeps a book (usually now on a computer) of planned, "elective" operations, giving details of the patient, the procedure to be performed and the surgeon.

Customarily, each procedure is scheduled for a particular time, and the patient before he enters hospital is given the day and the anticipated time of his operation. Usually the patient has his operation on the expected day. Often he does not have it at the exact time specified, because of various possibilities such as previous cases taking longer or shorter times than expected, or the need to perform emergency procedures, which of necessity have been given priority over your elective, non-emergency, operation.

Do I want him to be my surgeon?

At some point, before, during or after your interview (consultation) with this surgeon, you are likely going to find that you are asking yourself this question. You are after all committing yourself to a pretty important undertaking, routine to him almost certainly, but certainly not routine to you.

One hopes that there is no doubt in your mind. You have found him pleasant and straightforward, he listened to your questions and gave you simple and understandable answers.

But what if he didn't? What if he seemed like his mind was elsewhere, or he wasn't very skilful at explaining himself? Does this mean he should not be your surgeon?

It is in part because of situations of this nature that the system of being referred by your family doctor to this consultant has come about. It is to be hoped that you know your family doctor and trust his judgement. As in all walks of life, the man with the most charm is not necessarily the man with the most competence. It is to be hoped that your surgeon is film star material, tall, handsome, considerate and charming. I suppose there are some surgeons like that, but mostly they're plain looking guys who are just good at their job. And when I have an operation (yes, Virginia, surgeons do have operations!) *"good at his job"* is what matters to me most! "Charming" would be nice as well, but "good at his job" is really the only thing that truly matters to me.

You might find yourself tempted to ask the surgeon, "Have you ever done this before?" That's not a question I've had put to me often, in fact I can only remember once when it was about a very common operation which I had performed many hundreds of times. If you have doubt whether the surgeon you're talking to is qualified to undertake the operation you and he are discussing, perhaps it would have been better to have asked your family doctor before you went to see the surgeon.

Should I have a second opinion?

I believe that the concept of a "second opinion" originated from the times when doctors were a great deal less specialised, but often one doctor would know a colleague who had more experience in treating a condition than he had. This would be a situation where a general physician (now called family doctor) would be uncertain whether a patient with abdominal pain should make a difficult trip in the middle of winter to see a surgeon in a distant town.

In most parts of Canada such a situation no longer obtains, and it is the patient who talks of a "second opinion" and not too often the

94

doctor. If the doctor does need the advice of a colleague it is generally given on an informal (for which read no-charge) basis, what we generally call "coffee room" or "corridor" consultations. The treating doctor describes the essence of the problem to his colleague and gets his unrecorded (and unrewarded) advice. We do seek consultations on a formal basis from our colleagues. A surgeon may want an opinion on whether a patient needs special pre-operative lung treatment for instance, and may request a formal opinion from a lung specialist (respirologist). But this we call a "consultation" and do not class it as a "second opinion."

The patient asking for a "second opinion" in my experience has usually been prompted by some (interfering) relation who is sowing distrust in the mind of a previously perfect satisfied patient. The patient is perfectly entitled to further opinions if he wants them, and under the rather strange system we have in Canada he can go on asking for as many opinions as he pleases since it costs him nothing. My record to date is one young lady who had seen seven fully trained orthopaedic surgeons, for a perfectly straightforward condition, she had received exactly the same advice from all seven, but was still demanding further opinions.

The problem with a "second opinion" is that it may differ from the first, there is after all more than one way to skin a cat, and this difference may seem more important to the patient than to the involved surgeons. A difference of opinion does not make one right and the other wrong, think of getting travel directions. So what does the patient do? Why not ask for three, for four or five opinions?

When asked about this, unrelated to my own surgical practice, I have for years told patients that they must first find a surgeon they trust, listen to what he says, and do what he says. If you don't, for whatever reason, find you have confidence in the surgeon you're talking to, then find another one. If things don't work out after the surgery the way you had hoped and expected, it is by that time too late to wish you had gone elsewhere.

University hospitals

There are some particularly complicated operations, or operations that require unusual skills and equipment. These operations may be performed in only a few hospitals, and such hospitals are often associated with a medical school.

You should be aware that the function of medical schools is to teach doctors in their training, at all levels of training. You will therefore be asked the same questions, and undergo the same examination by a number of different persons. Although they are learning from you, they are not "practising" on you. At each level of training they are restricted in what they do to their level of established competence. The person who does your operation, however, may not be the senior consultant surgeon to whom you were initially directed.

POST-SURGERY ISSUES THAT REQUIRE YOUR PRE-SURGERY PLANNING

How long will I be in hospital

You will want to know how long it is expected you will be confined to the hospital. Will it be "day stay" of a few hours? Overnight? For several days?

Your surgeon will tell you his usual experience for the type of surgery under discussion, but remember always that the unexpected does happen, it happens to *somebody* and it may be your turn to be that somebody. You must allow for the unexpected in your plans.

Will treatment be concluded by this operation?

Although it is to be expected that your surgeon will make you aware of any significant factors in your post-surgery care, it is not unusual for a patient to wish to be sure whether they will be in need of:

another operation,

chemotherapy,

radiation therapy,

physiotherapy.

After the surgery:

Will I walk unassisted?

or will I need:

crutches, canes, or a walker?

someone to help me to get around?

Will I have casts, bandages, braces and so on?

what arrangements do I need to make?

How long will I be disabled after my return home:

from care of the house and yard?

from care of my family dependents?

from my job?

Your surgeon cannot answer any of these questions without having a good understanding of your responsibilities in the house, the yard, as a caregiver, and the physical nature of your work.

Issues to discuss with your employer

will your employer allow work modification on return?

shorter hours?

easier tasks?

change of position, for instance, sitting instead of standing?

for how long would he permit this?

if he would allow you a permanent job change,

would a permanent job change save you from having the operation?

While you're disabled who will care for:

the children?

elderly parents and significant other?

goldfish, cat, dog, hamster and other pets?

your home?

If disability continues longer than anticipated:

do you have a fail-safe plan?

could you cope with an extended loss of income?

Further physical examinations

Because of the need to economise in the use of hospital beds, the pre-operative medical examinations and the necessary completion of administration forms, is commonly now undertaken in a series of separate visits the patient is required to make to different offices, instead of all these persons coming to the patient after he has been admitted into the hospital.

Because of the long time delay between the initial consultation and the date the surgeon can obtain for your operation, it is quite possible that he will wish to see you again, in part to repeat the physical examination and determine that your condition has not significantly changed, and in part to be sure that you do not have any unanswered questions; we all think of questions we ought to have posed after we've gone home, or our wives think of them for us.

It is also possible that, because of your general medical condition, your surgeon may request a pre-operative consultation with a medical specialist (internist) in regard to your general health, or perhaps more specifically your heart (with a cardiologist) or your lungs (with a respirologist). These same specialists may assist in your general medical care in hospital after you have had your operation.

Some hospitals have standing rules that patients with weight problems, heart or chest problems, or other conditions which might pose a hazard or require particular attention during anaesthesia, will have a pre-operative consultation with an anaesthetist. In some larger Canadian hospitals this pre-anaesthetic consultation is conducted on an impersonal clinic basis, but at the time of writing, most consultations of this nature are given by the same anaesthetist who will look after you during your surgery.

Pre-admission teaching

Most hospitals have pre-admission clinics conducted by nursing and administrative staff, usually working on a full-time basis in this clinic and expert at answering the patient's questions. The clinics are commonly called "pre-admission teaching" and in hospital-speak PAT. The nurses will take time and trouble to explain the hospital routines to you, exactly what to expect and how everything is done. They will take time in answering your questions. It is my personal belief that fear is largely based on apprehension of the unknown, and that a great deal of the inevitable fear of surgery is allayed by these explanations, and by the friendly faces. Routine blood tests, and possibly a cardiogram, will be made by the appropriate technicians.

You will be registered in the hospital, and given a "swipe card" which is used at every stage in the hospital, just as a credit card would be used in making a series of purchases in a shopping mall. You should retain this card, and bring it with you every time you come back to this hospital. It has a permanent number on it, and (regretfully) it is now by our numbers that we are identified.

With the assistance of the nurses and the administration staff, you will complete a number of forms. They will need to see your insurance card (actually see it, not have you try and remember the numbers), as well as your hospital swipe card, if you already have one. If your operation is related to a workers' compensation claim, they will need the claim number. If you have an insurance scheme that will pay for private accommodation, they will need to see the card or documents relating to that. They will want to see your medications so they can list them, you will be asked again about allergies.

All of this is time-consuming and boring. They make it as inoffensive as possible, and you will, of course, bring all they ask and do all they request. You are at liberty to refuse, the hospital is at liberty to deny you your operation.

Medications to discuss with your surgeon
You will need to have a clear understanding with your surgeon whether to continue or discontinue your medications. This discussion preferably should be held some little time before the operation because there are certain medications that have highly undesirable effects during surgery, but these potential effects don't disappear the moment the medication is stopped.

Whether you should continue with these medications, is a decision you will make in conjunction with your surgeon and other medical advisers. Typical concerns that they will have will be in regard to aspirin which reduces the blood's clotting capacity, and is in common use for that express purpose; in regard to more specific "blood thinners" which do the same; birth control pills which may be associated with clots in the veins blocking lung function, especially in less young women and those who smoke; the use of cortisone (prednisone), even within the last few months, which may require special medication while you are under the anaesthetic; insulin and other medication taken for diabetes; medication taken for heart or lung problems.

Most of what has been discussed is in relation to discontinuing medication before surgery, but there are some special circumstances in which medication should be *started* before surgery. Some patients who have had chronic heart valve infections (sub acute bacterial endocarditis, SBE) and some patients who have had heart operations, are advised that they should take antibiotics in association with surgery. Some patients with recurring kidney and bladder infections are given the same advice. Your surgeon should be made aware of any problems of this nature.

These are the commonest drugs of concern, but you must be sure that **all medication** you take or have taken in the last few months is reported to your surgeon and to the nurses at the pre-admission clinic. **They are all trying to help you, ultimately it is you who are responsible for helping yourself.**

Insurance forms

Many employed persons have forms that need to be completed in order to draw sickness and accident benefits (S&A) or other types of temporary disability insurance when off work.

Don't expect the doctor's office to fill your form out while you wait, they cannot put aside all other activities, looking after other patients, to do your forms the moment you present them. These forms are often quite complicated and demanding in their requirements; if they are not fully and correctly completed your insurance carrier may arbitrarily refuse you any money; give the doctor's office enough time to do them right!

Do expect to be charged for this service, it is specifically not covered by provincial health insurance programmes as an insured service.

Immediately before the operation (pre-op)

As stated elsewhere, it was at one time the usual hospital routine to admit every patient to hospital at least one day before surgery. Undoubtedly this was to a large extent for the convenience of hospital personnel, and certainly made little sense to many healthy patients.

Because of budget pressures, most hospitals now do not bring the healthier patients in any sooner than necessary. This reduces the number of days the hospital beds are filled, resulting in most hospitals in a reduced number of beds and lay-off of nurses. It is in many ways more convenient for patients who have work to do and families to care for. It does result, however, in the need for patients to take responsibility for themselves, rather than being under the supervision of a nurse. In most instances this works out perfectly well, in a few instances because of wilful disobedience of the instructions given to them, the patient's surgery has to be cancelled.

Admission to hospital might be one day before, or several days before surgery, if extensive preparation ("work-up") is required. In most hospitals such patients, after being processed in the Admitting Department, will be taken directly to a surgical ward (also variously called a surgical floor or unit).

If it is anticipated that you, the patient, will not need to remain in hospital after the operation, you will usually have your administrative processing in what is commonly called a *short stay surgical unit*, different hospitals employ different terms.

If it is anticipated that you will be admitted to a surgical ward after your operation, but your general condition is such that you do not need to be in hospital one or more days before your operation, you will probably go initially to the short stay surgical unit for the admission processing, and then after completion of your operation you will be taken to the surgical ward, a process known in the hospital as *Same Day Admission* (SDA).

You will generally be given advice in your surgeon's office, by the surgeon, or more often by his nurse or his receptionist. In order that all patients have the same understanding of what will be required of them, most hospitals now provide the patient with brochures; your surgeon's office may have a supply of these from the hospital he works in. If you are not given one, ask if they are available.

The advice that you are likely to be given is as follows, but you must be certain for yourself that this is what is wanted by your hospital from you, there are always variations and exceptions.

Smoking

The first thing you should give up before surgery is the habit of smoking; we all know this is easier said than done, but as surgeons and anaesthetists we all know that you can tell the smoker from the non-smoker by listening to his chest with a stethoscope. The smoker's chest is "noisy." During the anaesthetic he is more likely to get into difficulties with his breathing, after the anaesthetic he is more prone to lung infections, such as pneumonia. Even if you can't bring yourself to cut cigarettes out altogether, do cut them down as much as possible, for your own sake.

Medication

It is usually advised that if you are taking insulin (which is used for the treatment of diabetes) that you take your usual dose the day before surgery, and do not take any on the day of your surgery. There may well be personal variation in advice on the use of insulin, and you should seek this. If you usually take pills to control your diabetes, you may be instructed not to take them on the morning of your operation.

It is usually advised that if you are taking pills for problems with your heart, with blood pressure, with asthma, or for seizures, that you take these on the day of your surgery, using as little water as possible to help you to swallow them. Puffers (inhalers) for asthma, should be used on the day of surgery as on any other day.

You will be advised that you should bring all your medications (meds) with you to the hospital.

Blood thinners, and aspirin which is also a blood thinner, are usually discontinued ten days before surgery. You should get specific advice on this and on whether you should discontinue anti-inflammatory medication such as cortisone (prednisone) and NSAIDs.

Eating and drinking

You may be advised multiple times that you should have nothing to eat or to drink after the midnight before your operation.

The reason for this advice is plain, and brutally simple. You may vomit the contents of your stomach when the anaesthetic is started, this acid and food may go down into your windpipe, into your lungs, you may choke and you may die. It doesn't happen often, but it does happen, and I've seen it.

It is hard to get some patients to understand that the "nothing after midnight" rule is for their benefit. They are the one who's going to die, not the surgeon or the anaesthetist!

We all take this rule seriously enough that if one of my patients has disobeyed it I invariably cancel a non-emergency operation. Most other surgeons and nearly all anaesthetists will do the same.

Your own property

Bring as little as possible in the way of personal possessions to the hospital. Hospitals are just as likely as other places to have dishonest persons in their building; they probably will not be hospital employees, but they might be other patients or their relatives. So the standard and fully intended advice is, "Leave your valuables at home."

If you wear rings, they are better left at home. If you cannot get them off, because of the possibility of the circulation in your fingers being cut off by the ring, hospital personnel may have to remove them before the surgery. This is particularly likely to be necessary if the arm on which you wear the ring is to be operated on.

Other personal jewellery, including the rings that are passed through various other parts of the body, should also be left at home.

Do bring with you your eye glasses, your contact lenses with container and solutions, your hearing aid, your crutches, your artificial leg, and any other device you normally use for communication with other persons or for walking. If you are under treatment for sleep apnoea, you should bring your machine with you. Because of concern about sleep apnoea after an anaesthetic, you may be required to remain overnight in the hospital for observation; this is a matter you should have discussed with your surgeon and your anaesthetist; they don't know you suffer from sleep apnoea if you haven't told them about it!

Your body

It shouldn't be necessary to tell a patient to wash himself before he comes to hospital, unfortunately not everyone does; I doubt whether anyone who reads this book will need to be told to bath or shower on the morning of the surgery. Some of the cleanest and most scrupulous patients do need to be told that they should not put on any make-up, that includes finger and toenail varnish as well as the face. Hospital personnel involved in your care will check the general appearance and colour of your eyes, cheeks, finger and toe nails, and need to have all areas of your body free of artificial materials. There's no reason why your friend or relation shouldn't bring your make-up for you to apply when you are discharged.

Pimples and nail infections fairly obviously mean skin infection. Not everyone realises that any break in the skin, any small cut, any abrasion, any insect or animal bite also means infection. Your surgeon should be made aware of any problem of this type before you enter hospital; he may consider that for your sake it may not be safe to proceed with the operation. Pimples on the face will probably not be considered a reason to cancel a hernia operation. But if you have an abrasion on the front of your knee this may very well be considered to pose too much risk to put an arthroscope through that area and into the joint, possibly resulting in a joint infection. If you have a low grade infection around a toe nail, this may very well be considered by your surgeon as causing too much risk for a replacement hip operation (total hip replacement, THR).

There are hidden infections that are just as important in deciding whether they are posing a risk; these are of particular importance in implant surgery, for instance heart valves or joint replacement. The commonest are chronic infections around the gums and teeth (pyorrhoea) and chronic low grade infections of the kidney and bladder (urinary tract infections, UTI). Your surgeon must be informed about this before any operation is arranged in order that treatment can be given, and possible disasters averted.

Administrative matters

If you are not fully competent in the working language of the hospital where you are being treated, you should bring someone with you to act as an interpreter. Most hospitals have a list of employees who are adequate to interpret in the many languages that are now spoken in Canada, but these are hospital employees, they are supposed to be at their work, and their services are made available only for emergencies.

All patients who are legally incompetent as a result of their mental status or their age must be accompanied by a legal guardian. This might mean your father who has Alzheimer's disease, or it might be your fifteen year old son.

Hospitals work under the strictest of legal controls; they will not put themselves at risk of being sued because it wasn't convenient to you to come with your relation. The surgery will be cancelled unless it is an absolute emergency.

Even though you showed your provincial health card when you came for your pre-operative assessment, bring it again. They are often used as "swipe cards" as well as bearing your number.

Why do they keep asking me for my name and why I'm here?

Once again, this is a hospital routine established and followed for your protection. Don't get irritated with the clerk or the nurse, they are doing the job they were told to do. You will have been given a plastic band around your wrist, with your name and other details; they will check not only what you say, but also the wrist band and their "job sheet."

Would you prefer to have your stomach removed instead of your hernia repaired, just because you were not the only John Smith going to the operating room that day?

Would you prefer to have your left knee operated on instead of your right, just because your surgeon's receptionist was confused about which side was to be done when your operation was booked in the hospital? Or perhaps you discussed both knees with the surgeon, he thought you wanted the right one done, you had in fact said you wanted the right one done, but now it's the left that's really hurting the more.

Mistakes happen in all walks of life. The more persons involved in the transmission of a message, the more likely there is to be a mistake. You can make an estimate of how careful the procedures are in this hospital where you are about to have your operation, by the number of times at every point of transfer someone checks who you are and what operation you expect to have.

Smile and thank them for being so careful!

On the way to the operating room
It used to be standard practice for the anaesthetist to order pre-operative sedation for the patient to "calm their nerves." This possibly also helped in the actual administration of the anaesthetic. Now as a

general rule the patient is not sedated prior to surgery; there are, in this as in all things medical, some exceptions. The anaesthetist may believe sedation is advisable for you for some particular reason. If you have strong feelings about this you are always free to ask for, or refuse, sedation.

One or more persons will check who you are, what is to be done to you, and by whom.

From the surgical ward or short stay unit you will be brought on your back on a wheeled stretcher to the operating room.

You are not likely to be given any option about the use of a stretcher. Hospital practice has not yet advance to permitting perfectly healthy young persons to walk, doubtless one day when the waste of hospital labour is evaluated, the pre-op patients will be allowed to walk!

In the operating room

The term "operating room" is used both for the whole suite of rooms, that is the entire area where surgery takes place, and for the individual room where one surgeon operates on one patient.

Once again, one or more persons will check who you are, what is to be done to you, and by whom.

You will wait in the corridor, on the stretcher, bored out of your skull, apprehensive, and supposing every casual remark made by surgeons and nurses as they walk up and down the corridor, is about you. Some hospitals are civilised enough that patients waiting for surgery have a special lounge, TV to watch, magazines to read. If your hospital's like this, be grateful, most are not.

After a ten minute wait in the corridor, which will seem like ten hours, a nurse will come, and once again, one or more persons will check who you are, what is to be done to you, and by whom. If you are the person expected, you and your stretcher will be wheeled into the actual operating room, where the staff will ask you to move from your stretcher and onto the operating table. There follows the usual confusion with sheets, blankets, operating gowns and so on.

When you have settled onto the operating table, once again, one or more persons will check who you are, what is to be done to you, and by whom.

At this stage you will have an excellent view of the ceiling and will notice various suspended lights; one or more nurses will be clattering surgical instruments on a table at the side of the room. Everyone but you wears a surgical mask. A person approaches you, usually from behind at the head of the table; you will know he's the anaesthetist because his mask doesn't cover his nose!

In the operating room, the anaesthetic

You may already have met the anaesthetist, he will probably introduce himself, and probably his first act will be to look at the band on your wrist, and verify verbally that you are the patient he expected, and that you yourself are expecting to have the operation that's scheduled on his list.

He will probably use his stethoscopes to listen to your heart and lungs. He may tape some electrodes to your chest; these are connected to a cardiac monitor which gives a continuous tracing of your heart's electrical activity, the typical up and down squiggle of the TV dramas. A metal clip, like a sponge rubber lined clothes pin may be put on a finger; this gives the anaesthetist a continuous read-out on the oxygen in your finger's arteries.

He is now in a position where he can watch (monitor) the function of your heart and lungs throughout the operation.

The next act will be to set up an intravenous infusion, you'll have seen this on TV if not in person. A needle is put into your vein, frequently a small needle into a vein at the back of the hand. It's connected by plastic tubing to a bottle slung from a pole, and that's usually the last unpleasant thing you remember if you have a complete (general) anaesthetic. Other forms of anaesthetic will be described below.

The days when the surgeon had his chauffeur give the anaesthetic are long passed. Giving an anaesthetic by dripping ether or chloroform onto a gauze pad held over your nose and mouth are gone, but not so long ago; I remember many anaesthetics I gave like this, but my younger colleagues have never seen ether or chloroform in use. Like everything else in medicine, anaesthetics have gone *"high tec."* Of course this means *high tec* equipment, highly trained and specialised personnel, lots of different medicines and gases to choose from, and lots of expense!

Usually the anaesthetist will inject one or more drugs into the tubing of the intravenous (IV), you may feel a cold or warm sensation in the vein, but not any real discomfort. The method varies from doctor to doctor and with the patient's individual needs, but standardly one of the substances injected will cause loss of consciousness; one may also be used to paralyse the muscles. The anaesthetist at this stage will usually put a mask on the patient's face and squeeze oxygen ("bag the patient") into the lungs; some anaesthetists have the patient wearing the mask before commencing the anaesthetic (induction) in order to get a maximum oxygen content into the blood at the outset.

Depending on the case, the anaesthetist may continue to bag the patient, who is also breathing for himself, or may pass a tube (intubate) into the windpipe (trachea) and allow breathing to be performed automatically by a machine; there are many variations in between. In general, short anaesthetics are carried out without intubation, and longer anaesthetics or patients with short thick necks are intubated.

The anaesthetist decides how he will administer and conduct the anaesthetic. There might be some cases where this is discussed with the surgeon, or probably a consulting respirologist, but the decision is made by the anaesthetist and the surgeon rarely has any say in this, any more than the anaesthetist tells the surgeon how to do his operation!

The above description is relevant to a complete (general) anaesthetic given in association with breathing different gases (inhalation). Other forms of anaesthetic given for the relief of pain (analgesia) in only part of the body (regional anaesthetic) are in common use. As stated already, the decision in regard to what type of anaesthetic will be given is made by the anaesthetist. Where regional anaesthesia is concerned the surgeon is likely to have more input into the decision, but the ultimate decision is still made by the man who does the job, the anaesthetist.

Most persons are familiar with the dentist freezing the mouth with local anaesthetic, thereby making an operation on a tooth painless even if it's not altogether pleasant.

The anaesthetist, or sometimes the surgeon himself, may perform the same type of procedure on the arm or the leg, injecting a nerve or a combination of nerves, rendering a particular area of a limb without sensation (local or regional anaesthesia).

The same type of "freezing" drug may be injected into the spinal canal, mixing with the liquid (cerebro-spinal fluid, CSF) that flows from the brain and around the spinal canal and nerve roots. The nerve roots are temporarily rendered without function, and the lower part of the torso and the limbs can be operated on without causing the patient any pain.

A similar procedure is performed on pregnant women, but in their case the freezing medication (local anaesthetic) is usually injected outside the tube (dura mater) that contains the CSF. This type of regional anaesthetic is known as an epidural.

Throughout the anaesthetic your heart and lung function will be monitored using the continuous EKG cardiogram, and the oxygen monitoring device clipped on to your finger. There are warning bells on most of these machines should anything untoward occur, and all the medications and equipment that might be needed to deal with an unexpected occurrence is to hand. In most operating room suites when the unexpected happens, and it does from time to time when least anticipated, there is another anaesthetist available to give your anaesthetist any assistance he might need.

Like crossing the road, anaesthesia is not without risk, but your anaesthetist's training has been concentrated on what to do when the unexpected happens.

In the operating room, the surgery

We are not in this book going to deal with all the techniques of all the different operations that are carried out, it would be far too space consuming to describe all the operations that exist and rather pointless when you are only going to have one of them. I find in fact that most patients aren't really that interested in the operation details.

If you are, then ask your surgeon to explain to you what he is going to do. Don't ask a friend, don't watch it on TV, don't ask a nurse, none of these are likely to be what your own surgeon is going to do.

AFTER THE OPERATION

Where do they take me next?

Most patients are taken to the recovery room before they are fully conscious; their stretcher is pushed/pulled by the anaesthetist and one of the operating room nurses. For the first time the responsibility for your care will be transferred without anyone asking you to identify yourself. The anaesthetist and the operating room nurse will give a brief report to the recovery room nurse, check all is in order and then leave.

You may be breathing without any assistance before you are transferred from the operating room table and back on to the stretcher. You may still be intubated, or you may be alert and fully aware of what is going on around you, or you may be muzzy, or you may be awake and talking but remember nothing of what you've been told. Contrary to popular rumour, I have yet to meet a patient who told all the intimate details of their life history when coming out of the anaesthetic!

You will remain with the recovery room nurses until your condition is completely satisfactory. They will check your pulse and blood pressure at intervals, and will look at your bandages or dressings; they are not likely to change them. When they are satisfied with your condition you will be transferred, still on a stretcher, to the next phase of your care. If you are not to be admitted this will be back to the short stay surgical unit, and you will be discharged from there. If you are to be admitted, you will be staying in the hospital overnight or longer, you will be taken to the surgical ward.

The exception to this direct transfer from operating room to recovery room is the patient who goes to the intensive care unit. There are certain operations which require the patient to be more closely monitored, with more technical equipment than is available in the recovery room; for these patients direct admission to the intensive care unit (ICU) is arranged prior to their operation.

What do they do about my pain?

Orders are routinely written by your surgeon or his assistant to ensure that you have adequate medication for pain relief after surgery. The nurses have a schedule for progressive administration of pain relieving drugs (analgesics) during the period of gradual recovery from the effects of the anaesthetic. This is one of many reasons you will have been asked about allergy to medications, when you're wanting medication for relief from pain, or after it's just been given to you, is not the moment to remember that the last time you took Demerol you vomited for two days!

There is now an apparatus, in common use, called "patient controlled analgesia" (PCA). This is effectively a pump, clamped to your IV pole, which permits you to administer the pain relieving medication yourself, on a continuing basis and under your own control, rather than watch that clock on the wall for the hand to come round to when you can have your next needle.

We haven't yet learnt how to abolish pain, but great strides have been made towards relieving it.

What to expect if you are admitted to hospital after your surgery

Admission to hospital before any but the smallest of surgical procedures, and a stay of several days in hospital after an anaesthetic was the standard practice until a few years ago. The patient might have asked, "Why do I need to stay in hospital?" and the standard answer was the usual, "Because that's the way it's always been."

Most of us in the era before universal insurance recognised that the patient, who had for financial reasons to get up and get going, often did much better and had fewer complications than the patient who was kept on complete bed rest for several weeks, with family and private nurses running attendance.

Two factors have come together in recent years, one is purely related to money, the other is related to good patient care.

The "good patient care" part has just been described, the sooner a patient gets going after his surgery, the less time there is for physical and moral deterioration. Bodies lose strength and substance if not used. Minds become dependent. The basis of "rehabilitation" is the earliest possible restoration of the person to their pre-surgical, pre-injury, status.

The other factor is the impact of insurance on medical care. In the days before universal health care hospitals were funded by charity or by charging the patient for the services they received. Hospitals are now funded out of government budgets which are tightly controlled and continuously reduced. Hospital administration has therefore no choice but to cut back on the services provided. Since 75% of average hospital costs come from staff payroll, cutting staff is the first economy. Cutting the number of beds and reducing the hours the operating room can be used is the result.

All of this is a long way round to approaching the usual question, "How long will I be kept in the hospital?" The answer is very simple, "No longer than absolutely necessary for your recovery."

Here we emphasise "recovery." Because of economy measures hospitals can no longer be used as hotels. There are patients who have very little wrong with them who enjoy being in hospital and the attention they receive from staff and visitors. Hospital budgets no longer leave this as a possibility.

While in hospital you will receive all the care your condition requires, and when your condition no longer requires that you receive your treatment in hospital, when you have reached a point that it could just as effectively be carried on outside a hospital, you will be asked to leave, for which doctor and hospital-talk is "discharged."

While I'm in hospital, who else will treat me?

The persons most immediately responsible for your care are the nurses.

It isn't always easy to recognise who is and who isn't a nurse since they have abandoned the traditional white starched uniforms with caps, and wear more "customer friendly" coloured clothing, which is in general more comfortable for them as well as their patients. The nurses will introduce themselves to you. However, since through each twenty-four hour day there are usually three shifts, and since few nurses work more than five shifts a week, and some only two, don't expect that your continuing care will be given by only one nurse.

Physiotherapists may be involved in your care if you have had an operation that requires an exercise programme. They provide care not only for patients who have had spine and limb surgery, but also for patients who have had other procedures and need an exercise programme to maintain or regain body strength and mobility.

Respiratory therapists may work with you if you have need of their skills and equipment to improve your lung function. There are numerous other specially skilled persons, variously called therapists or technicians, who will assist you in your post-surgery care, according to your particular needs.

The attention of all of these staff members will be directed at your rehabilitation. Depending how you view this yourself, rehabilitation can be thought of as getting you back to health as soon as possible, or getting you out of hospital as soon as possible. We all hope you take the first viewpoint!

To the outer hospital door

It is a rule in every hospital I know that the patient is to be taken by a member of the hospital staff in a wheelchair to the outer hospital door. The explanation lies in the hospital's legal responsibility for your safety. Although you may feel just fine, you may look just fine, and you may feel foolish being pushed in a wheelchair by someone twice your age, every now and again a patient faints or trips and the hospital may be sued. So, they just aren't about to take that risk.

Transportation from hospital to home

The transfer of responsibility for your safety is taken very seriously by all hospitals. They require that a "responsible person" present and identify themselves to the nurses who have looked after you before your discharge, and that this responsible person will accompany you in a taxi or drive you home. **You will be expressly forbidden to drive any vehicle yourself.**

When at home, a responsible person must be with you, at least for the first night.

These rules are developed from long experience by hospital medical, nursing and administrative staff. You will, as your side of the bargain, be expected to comply with them.

Discharge instructions

The nurse who cares for you at the time of discharge will check whether you have or need a prescription for medication to control post-operative pain, and prescriptions or an adequate supply for the other medications you regularly take. She will also check whether you have, and understand, arrangements for further care from your surgeon or other health care provider, and will ensure that you know where to turn to get help should you need it.

Many hospitals supply the patient with a sheet of post-discharge instructions and advice, with specific comments on the medication and post-operative instructions for you, recognising the frailty of person's memories.

Rehabilitation hospitals

There are a few hospitals which specialise in rehabilitation, the restoration of your body and mind to the level of maximum possible improvement. You may be discharged from the hospital where you had your surgery, and admitted to a rehabilitation hospital. Some general hospitals have rehabilitation sections, variously called wings, units, floors and so on. They often constitute a separate hospital within a hospital which is of consequence to the administrators in devising their budgets, but not of consequence to the patient.

If you are transferred to such a place, this does have a special benefit from your point of view in that it gives you an institution which specialises in rehabilitation, exactly what you need at this stage.

The patient with a large wound

If a large incision has been necessary, for instance because of removal of an internal organ involved in cancer, there may very well be the need for changing of dressings on a daily or more frequent basis. It was at one time the practice to keep the patient in hospital until all dressing changes were completed, stitches had been removed, and often the patient was kept continuously in bed until that time. At this point he was physically weak and emotionally dispirited. For all the reasons previously given, this is no longer the way patients are treated after surgery in Canada.

126

Long before there was any provincial health insurance, Canada had the services of nursing organisations that provided care of the patient in their homes. These organisations such as the Victoria Order of Nurses (VON) continue in this task, but have been supplemented by several others, working in the field that is now called "home care." This whole field is in a state of flux, fortunately flowing towards an increase in service to balance the decreased services the hospitals are permitted to provide. There is in consequence a wide difference in availability of home care services comparing province with province and different regions within any one province. Currently, in Ontario home care services are limited to 90 hours a month, if more care than that is required, admission to a nursing home may be indicated.

Most hospitals now have a person who is variously designated as "discharge co-ordinator" or "home care planner," who will, in conjunction with your surgeon and family physician arrange the provision of needed home care services. I emphasise "needed" because the services are costly, budgets are limited, and the co-ordinators and administrators must see that their services are provided to patients and families who need them, and are not used up by patients with little or no need of them.

Among the potential services that may be provided by a home care organisation are:

nurses to do dressing changes, remove sutures, and supervise the recovery process;

personal care attendants who will help with washing and grooming;

physiotherapists who will plan and supervise exercise programmes;

housekeepers who will provide light housekeeping services, shopping, and light cooking.

If you are one of the fortunate few whose family physician makes house calls, he may also visit, examine and prescribe for you at home. However, the more usual pattern is for medication to be ordered over the phone, and the family physician not to see you until you are fit and able to come to his office.

Visits made after the operation to your surgeon might be in his office, or might be arranged in the ambulant care section of the hospital, the same area that used to be called the out-patient department (OPD). It would be unusual for a surgeon to visit you in your own home.

The patient with a small wound

Having emphasised in previous sections that "minor surgery is what other people have," it is fair to say that there is such a thing as "minor need" *after* surgery. For the average patient who has had an arthroscopy or laparoscopy there is very little post-surgical care required. You will need to visit your surgeon about a week after the surgery to have any stitches removed, and to discuss the significance of the operative findings. Other forms of smaller surgery requiring little post-operative care are usually followed by a single visit to the surgeon so that he may be sure there are no problems; when he is satisfied with your progress your family physician will continue supervision of your care.

Occasionally one finds a patient who believes the surgeon "lost interest in me" because he doesn't offer multiple visits to his office after the operation. This is based on a misunderstanding of the surgeon's role. The surgeon, like other consultants, provides a form of care that the family physician believes the surgeon can manage better than the family physician himself; for the remainder of your care, the family physician is the one you should look to.

WHAT IF IT WENT WRONG?

It has previously been emphasised several times that there is no such thing as a "risk free" activity of any kind, and that includes surgery. You must therefore take the possibility of a post-operative complication into account when calculating time away from the responsibilities of your job, house, yard, children, and other family members.

Complications shortly after surgery

If within a few days after discharge from the hospital some unexpected occurrence is found, for instance bleeding into your dressing, a fever, an increase in pain, it is usually better to contact your surgeon about this rather than your family physician. The family physician may have little understanding of the cause of your problem, not because he is medically ignorant but because he is at this stage not fully informed about what has happened to you.

The surgeon quite probably will not be immediately available to you, but his nurse or secretary can advise you; you can be reasonably sure that, although complications are now uncommon, yours will not have been the first one.

As a less satisfactory alternative you could go to the emergency department of the hospital where you had your operation. There will be a nurse on duty who has a full supply of surgical dressings available to her; probably there will be a doctor who can give advice; they will have a surgeon on call for advice if needed and if your own surgeon is not available.

If you aren't satisfied with the outcome of the surgery
If things haven't turned out quite as expected, and short of some very obvious complication such as an infection or bleeding, the first question you should put to yourself is, "Have I allowed enough time to get better?"

Everyone wants an instant cure, no matter how long it took for them to decide the problem needed attention. Improvement in function or reduction of pain after an operation often does take some time, it may take longer than either you or your surgeon had anticipated. So, rule number one: allow enough time for recovery.

If you have allowed plenty of time, and things have not improved, you should discuss this in a simple, direct, and non-hostile manner with your surgeon. He may have an explanation why it is taking longer than expected to get better; he may have a supplementary treatment he thinks would help.

You can of course discuss the failure of anticipated recovery with your family physician, but unless there are quite unusual reasons for doing so, I would strongly recommend against going to another surgeon at this stage to ask his opinion on why you haven't got better.

Should you be in the unfortunate, and I believe uncommon, position that you do not find you are getting the support you need from your surgeon, then I suggest discussing this with your family physician and that you follow his advice.

Don't go to your golf club buddy for his opinion!

www.ingramcontent.com/pod-product-compliance
Lightning Source LLC
Chambersburg PA
CBHW022003170526
45157CB00003B/1118